Contents

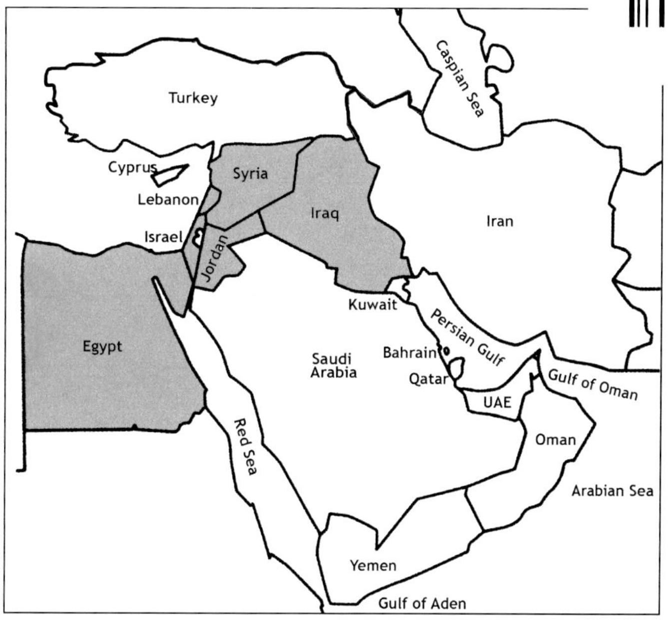

Helion & Company Limited
Unit 8 Amherst Business Centre, Budbrooke Road, Warwick CV34 5WE, England
Tel. 01926 499 619
Email: info@helion.co.uk Website: www.helion.co.uk Twitter: @helionbooks Visit our blog http://blog.helion.co.uk/

Published by Helion & Company 2025
Designed and typeset by Mach 3 Solutions (www.mach3solutions.co.uk)
Cover designed by Paul Hewitt, Battlefield Design (www.battlefield-design.co.uk)

ISBN 978-1-804515-73-0

British Library Cataloguing-in-Publication Data.
A catalogue record for this book is available from the British Library.

DEDICATION

For Air Commodore Abd al-Moneim Miqaati (8 July 1904–5 April 1982)
in memory of a meeting in Groppi's in 1970.

'"If men would behave decently the world would be decent" is not such a platitude as it sounds.'
'Charles Dickens', essay by George Orwell in *Inside the Whale and Other Essays* (London, 11 March 1940).

NOTES:
References to 'the author' in connection with interviews, correspondence and other forms of personal communication refer to Dr. David Nicolle. All the times given to Middle Eastern radio broadcasts as transcribed by the BBC Monitoring Service are GMT.

ETHNIC CLEANSING, WAR, TRUCE, AND ARMS EMBARGOES

Following the months of low-intensity warfare that erupted in 1947, accompanied by bouts of ethnic cleansing of the Palestinian population by Zionists, and then four weeks of intensive warfare during the second half of May and early June 1948, and largely thanks to tireless efforts by UN mediators, all directly involved parties agreed to a ceasefire. This came into effect on 11 June 1948, and would last for twenty-eight days. By and large the Arabs abided by its terms, whereas Israel violated the cease-fire in both spirit and deed. Although most of the Arab states had opposed an acceptance of this truce for fear of breaking the perceived momentum in their operations, intense diplomatic and political pressure from Britain, coupled with King Abdullah's insistence that the ceasefire be accepted, meant that the Arab League finally agreed. In fact, the Transjordanian Prime Minister, Tawfiq Abu al-Huda, threatened that his country would not only pull out of the war but would withdraw from the Arab League if the UN ceasefire were not accepted.

During this period both sides tried to reorganise and reinforce their troops, replenish their supplies and improve their equipment. However, the Israelis were able to take a much greater advantage than the Arabs, most notably by bringing in new material and men from outside the country. The following weeks were also a period of intensive activity for the conflicting parties as they tried to win political support at the UN. In the meanwhile, Count Bernadotte, the United Nations Mediator, was supported in his efforts to supervise the cease-fire by teams of Truce Observers formed of army officers from Belgium, France, Sweden and the United States.

Although generally considered as erupting on 15 May 1948, with an 'invasion' by neighbouring countries on Israel, the war in Palestine was already in full swing in November-December 1947. Not only in form of the first organised operations of ethnic cleansing of the local population by Zionists, but also in the form of terrorist attacks, like on this train that was de-railed by Haganah. (Albert Grandolini Collection)

LONDON'S ARMS EMBARGO

The Arabs lacked domestic arms industries and had neither the means nor the desire to draw in significant numbers of foreign volunteers. They also lacked contacts behind the 'Iron Curtain', while the UK, which had previously supplied many of the Arab regular armies, now blocked even previously agreed arms sales. Consequently, the Arab states could only improvise in their attempts to arm themselves, often by knocking on other doors such as the European black market. In the event, their efforts produced poor results – with the exception of Egypt's purchase of fighter aeroplanes from Italy. In addition to their shortages of arms and ammunition, the Arab armies lacked properly trained reserves to reinforce existing troops. In practice they could only reshuffle available forces, a significant proportion of which remained tied to internal security tasks at home. Nevertheless, the Arabs did add some strength to their armies in Palestine.

The records of an Arab Chiefs of Staff meeting in Cairo, on 15 and 16 June 1948, highlight the defects of the Arab armies at this time. After the UN arms embargo came into effect, and already having inadequate reserves of military equipment, the Arabs had run dangerously low on ammunition, spare parts and other essential material. Thus, by the end of the first four weeks of fighting, Arab forces suffered growing shortages of weapons, aircraft, and other war material. The shortage of ammunition involved all types, but especially shells for artillery, bombs for mortars, and shells, bullets and bombs for aircraft. Insufficient reserves to replace those of exhausted frontline units meant they could barely resume offensive operations. In fact, the Arab Chiefs of Staff concluded that their armies were not in a position to resume fighting, whatever their political leaders might desire. Thus, virtually all senior officers favoured a renewal of the UN Truce and went as far as to state – in writing – that their forces had reached the limit of their ability to advance.

On 16 June, at their meeting with Abd al-Rahman Hassan Azzam Pasha, the first Secretary-General of the Arab League (in office from March 1945 to September 1952), the Arab Chiefs of Staff explained the military situation on all fronts, highlighting the shortages and emphasising the need to deal with such problems before any resumption of fighting. For his part, Azzam Pasha asserted that a rejection of a continuation of the Truce did not necessarily mean a resumption of the military offensive. The Arab armies were, he suggested, to assume a defensive strategy but that '*we have to resume fighting if we are attacked*'. At the same time, the Arab League tried to allocate quotas for the distribution of available war equipment to each member state, though these quotas cited only quantities, not the quality of the material in question.

Generally speaking, the embargo worked to the advantage of Israel because it was much more effective where the largely British supplied Arab armies were concerned. In contrast, the theoretical embargo on military supplies to the Israelis from the USA and elsewhere was much less rigid. In fact, both Egypt and the Transjordanian Arab Legion's British commander appealed to London for desperately needed arms, but were refused. When Air Commodore Hayes, the British Air Attaché in Cairo, wrote his assessment of the REAF's role in the Palestine War (as a supplement to his First Annual Report on the REAF, written months after the fighting had actually stopped) he noted that the embargo on British military supplies had so handicapped the Egyptians that they turned to other countries for such purchases. One of the few that seemed prepared to defy the embargo was Italy, which was both willing and able to supply the REAF with combat aircraft. In fact, Italy was eager to find markets for its battered aero-industry as it emerged from the effects of the Second World War. The Egyptians' ability to partly resupply their forces with war material from British and American scrap dumps in Egypt, in addition to limited quantities of ammunition from stores abandoned in the Egyptian Western Desert since the Second World War, was a completely inadequate stopgap measure.

1

SECOND HALF OF THE FIRST PHASE (1 TO 11 JUNE 1948)

As the war effort on the Zionist side continued experiencing ever better coordination, the Israeli High Command decided to remove the Iraqi threat to Lajjun. For this purpose, it prepared a counter-offensive in the so-called Arab Triangle area. This was initiated on 1 June, with its first objective being Jinin. This Israeli plan called for a coordinated attack by the Golani and Carmeli Brigades, while the Alexandroni would mount a diversionary attack to split Iraqi forces. Accordingly, two Golani battalions prepared for the main thrust by seizing Arab villages in the Gilboa Mountains and near the Jinin to Afula road. At the end of this phase, the Carmeli Brigade would take control of Jinin itself. Meanwhile, the Alexandroni diversion through Wadi Ara threatened Tulkarm further divided the Iraqi forces.

1 JUNE
GROUND OPERATIONS

The Haganah High Command also approved Colonel Marcus' recommendations that an alternate route to Jerusalem be constructed south of the main road, from Dayr Muhaysin to Bab al-Wad and bypassing Latrun, which was firmly held by the Arab Legion. All available earth-moving vehicles, engineers, labourers and soldiers were therefore assembled to construct this road, or, more correctly, this improved track, under cover of darkness. It had to be completed before 11 June when the First United Nations Truce was due to come into effect.

Previously, the Israeli High Command had also decided upon a major assault to destroy Egyptian Army concentrations at Isdud, removing the threat to Tel Aviv. It was supposed to take place during the night of 1-2 June, but had to be postponed for 24 hours because of confused communications. What followed would become the week-long Battle of Isdud.

AIR OPERATIONS

Most of the Israeli Defence Force/Air Force's (IDF/AF; colloquially *Chel Ha'Avir* in Hebrew) offensive sorties were still carried out under cover of darkness, and at 0325 hours on 1 June, the airport outside the Transjordanian capital Amman was attacked by three Israeli aircraft. Two British Ansons of the RAF Levant Communications Flight were damaged, along with the Flight's offices and technical

A Palestinian girl fleeing a town shelled by Zionists, early during the war of 1948. Notable in the rear is at least one soldier of the Arab Legion. (Albert Grandolini Collection)

Hafiz Barakat, one of the leaders of Palestinian irregulars, as seen outside Hebron in 1948. (Albert Grandolini Collection)

stores. The Israeli decision to attack Amman at that particular time was probably because Arab League delegates were due to meet in the Transjordanian capital the following evening. An Israeli Bonanza was followed by an Argus then a Dragon Rapide, all being met by light anti-aircraft fire on their arrival. No casualties were suffered at the airport, but Israeli bombs killed six civilians and injured eight others in Amman itself. The Israeli aircrews did not report seeing any other machines in the air, but Baghdad Radio later claimed that an Iraqi fighter – presumably a Gladiator from Mafraq – had shot down a 'Jewish bomber' over Jericho. In fact, the Israelis suffered no losses and it is possible that this Iraqi sortie was entirely fictional. So ended the first Israeli air attack upon an Arab capital. Meanwhile, the Old City of Jerusalem and its sacred sites had so far escaped, despite a reported attack or attacks on its outskirts or outlying villages.

The British were furious about this assault upon the Kingdom of Transjordan, their close ally and virtually a client-state. In fact, the British Consul General in Haifa, the highest British representative in Israel, warned the Israeli authorities that if RAF instillations or aircraft were damaged in any future attack: 'we would be bound to defend ourselves and attack Jewish aircraft on the ground or in the air'. British units in Transjordan were also put on alert. The RAF even tried to arm the Percival Proctors of its Communications Flight at Zarqa and convert them into 'fighters' to protect both RAF Zarqa and RAF Amman. The size of the machines' fuel tanks was reduced and all unnecessary equipment was removed, while the man in the rear seat was to be given a hand-held Thomson submachine gun. Apparently, the RAF Communications Flight's CO, Flight Lieutenant J.H. Lewis, believed that such a modified Proctor could take on any Israeli Rapide which threatened the RAF airbases in Transjordan. Later rumours suggested that this 'high speed Proctor' would have a pair of machine guns mounted beneath its wings, and that the cockpit would be modified to convert the machine into a single-seater. Whether any such work was started, still less completed, remains unknown. Lacking an air force of his own, King Abdullah asked the Egyptians to retaliate for the Israeli attack. Abdullah was also more determined than ever to have his own air force.

Whatever was happening over Transjordan, the British remained extremely unwilling to supply Egypt with military equipment. In fact, secret Air Ministry cypher messages dated 8 and 9 June 1948 referred to material set aside for delivery to the Royal Egyptian Air Force (REAF) remaining at No.107 MU at Kafarit because transfers to the Egyptians were suspended indefinitely on 1 June 1948. This statement could, of course, be interpreted as indicating that some material had continued to be transferred to the REAF prior to that date.

Tuesday 1 June proved to be a busy day for the REAF, with the Egyptians responding promptly to King Abdullah of Transjordan's request for retaliation. According to Egyptian Communiqué No.21, broadcast at 2200 hours GMT:

> Our aircraft attacked a convoy of armoured cars north of Qastina, bombs and machine-guns destroying a large number. Panic-stricken crews abandoned their cars and fled. The Air Force made a concentrated attack on Tel Aviv, where our

King Abdullah with RAF officers at the RAF Amman, in June 1948. (David Nicolle collection)

> heavy and light bombers dropped masses of high explosive and incendiary bombs. Several explosions occurred in the area of the iron factories at Nakhlet Ganim (unidentified), which caused the main sections of the factories to collapse.

REAF aircraft also attacked enemy positions at Khulda and Be'er Tuvia, while reporting that fires could still be seen at Aqir aerodrome as a result of the previous day's raids. The Egyptians also suffered no loss.

For the Egyptian pilots held prisoner just north of Tel Aviv, such REAF raids were a source of considerable satisfaction, as Abd al-Rahman Inan recalled – though with some exaggeration:

> The Egypt Air Force's raids had increased over Tel Aviv day after day and become ten times a day. Some air strikes happened so suddenly than the siren started after bombs had hit their targets … Once I saw, between the camp and the sea, a soldier shoot a colour cartridge (Verey Light) when he saw aircraft before the siren announced an attack. The (Israeli) air defences consisted of Bofors and Bren guns on houses and in camps, with one large anti-aircraft gun – maybe 3 inch – which was between Jaffa and Tel Aviv. I think their purpose was not to hit aircraft but to raise the peoples' spirits … Yona camp was hit by bombs before I went there and the destroyed buildings and three cars were still there when we were moved to another prison. During one of the air strikes on the (Tel Aviv) port, one of (the) pilots saw the camp and strafed it with his guns. I thought my cell was his target because of the noise of guns above my head. Our building was weak, the wall being one brick thick and the roof made of wood. When our aircraft reached Tel Aviv all the (Israeli) soldiers went into shelters, leaving us three pilots.

At 1630 hours GMT the Voice of Israel claimed that: 'Enemy aircraft bombed Nir Am and Kfar Darom in the Negev; the aircraft met heavy fire from Haganah Army units, one of the enemy aircraft being brought down in flames near Sumsum, while another was hit over Dorot and had to make a forced landing near Gaza'. At 1840 hours GMT the previous claims appear to have been modified, with an Israeli Army Communiqué quoted by the Voice of Israel merely stating that: 'Enemy aircraft dropped bombs today in the Petah Tikva district and planes of the Israel Air Force drove off the attackers'. Further confusion has stemmed from the incorrect information that REAF Squadron Leader Muhammad Abu Zaid claimed to have shot down an Israeli aircraft on 1 June, whereas this claim was actually made on a later date.

At 1840 hours the Voice of Israel broadcast that: 'Three enemy aircraft attacked al-Afulah today'. This was within the Royal Iraqi Air Force's (RIrAF's) zone of responsibility and an Iraqi communiqué, broadcast by the Arab News Agency at 0900 hours GMT the following day, probably referred to Iraqi air operations on 2 June:

> The Iraqi Air Force carried out a heavy attack on Petah Tikva, east of Tel Aviv, causing big explosions and fires. The airfield

Squadron Leader Mustafa Abu Zaid with colleagues, at al-Arish. (Abu Zaid collection)

was also attacked, as well as the village of Lajjun, which was set on fire. Direct hits were registered on the town of al-Afulah, including the railway station. Big fires were observed after the raid. Aircraft also attacked the settlements of Hulda and Abu Shusha, on the main Jerusalem-Tel Aviv road. The enemy suffered great losses.

Egyptian Army troops in Palestine, mounting an armoured truck captured from the Israelis. (David Nicolle collection)

Syrian Communiqué No.18, broadcast from Damascus at 2050 hours GMT on the 2nd, similarly claimed that: 'A squadron of our Air Force continued its raids against Nahariya (on the coast) where a considerable part of the warehouses and quays were damaged and two vessels were sunk. Our squadron left their target with columns of flames rising from it. Another squadron attacked the Jewish colonies of Dan and Kfar Giladi (close to the Palestinian frontier with Lebanon) where the targets were heavily bombed'. The Syrian claims concerning Nahariya would be repeated in the October 1948 edition of the American magazine *Flying*, in an article by John C. Ross entitled 'Air War in Palestine'. Israeli aircraft were similarly in action over northern Palestine, the Israel Army Communiqué broadcast by the Voice of Israel at 1840 hours claiming that: 'Israeli aircraft bombed enemy concentrations and boats at al-Buteyha, where the Jordan flows into the Sea of Galilee'.

At 1645 hours GMT the following day the Voice of Israel added more details, stating that: 'Enemy artillery in the Tulkarm sector (Iraqi guns east of Netanya), which had been shelling nearby Jewish settlements, was attacked yesterday by our Air Force and artillery'. This broadcast also reported that: 'Jewish forces made progress in the Latrun district on the previous day (1 June), the Air Force raiding enemy concentrations in the vicinity'. A little later another Israeli broadcast included further extracts from an official communiqué to the effect that: 'An enemy plane dropped two bombs in the Kfar

Giladi area (at the tip of Finger of Galilee) last night. The Israeli Air Force attacked enemy concentrations at Jinin last night … Last night our forces advanced to within 5km of Tulkarm, which was bombed'.

2 JUNE 1948
GROUND OPERATIONS

Satisfied that General Naguib could hold the line at Isdud, General al-Muwawi returned to mopping up Jewish settlements that had been bypassed during the initial Egyptian advance. He began with Negba, lying east of al-Majdal, against which he sent his 1st Battalion. Negba was overlooked by various Arab-held hills, as well as by the major Egyptian strongpoint at Iraq Suwaidan. Egyptian commanders, having learned from previous experience, ordered a preliminary artillery bombardment, firing up to 600 shells per hour, plus an aerial bombardment. These were closely followed by an armoured attack, which almost overran Negba before being forced to retire. Nevertheless, Egyptian troops claimed to still hold some of Negba's outlying positions where, according Egyptian Communiqué No.22, they foiled Israeli attempts to retake lost ground. In the words of this report: 'Our forward forces repelled this attack with the aid of the artillery and the Air Force, inflicting considerable losses on the enemy who retreated in disorder … Our aircraft observed a concentration of armoured cars north of the Negba area and strafed it with gunfire and bombs, registering direct hits'.

A more objective analysis of this assault shows that on 2 June the first full-scale Egyptian attack on Negba set off from al-Majdal. The artillery and aerial bombardment was closely followed by three infantry columns, each led by one or more Sherman or Matilda tanks. The central column included seven assorted armoured vehicles and approached from the south-west at around 0700 hours. The leading tank destroyed one defence position before itself being damaged and obliged to withdraw, reversing to safety followed by the other vehicles. A second tank was then set on fire by Molotov cocktails and when two of the crew clambered out, they were promptly shot. Meanwhile an Israeli unit mounted in jeeps from the Negev Brigade, which had been attached to the Givati Brigade, appeared on the Julis hills to the north and opened flanking fire with heavy machine-guns. When Israeli 25-pounders joined in from the area of Kfar Warburg, the Egyptians retreated. Later that morning they laid smoke screens under the cover of which they evacuated their wounded and retrieved some damaged vehicles, including two half-tracks.

A covering force, reportedly consisting largely of local irregulars, was left to watch a now battered Negba while the bulk of the Egyptian force went on to capture the settlement of Nizzanim, between al-Majdal and Isdud. This was carried out in a combined attack by artillery, armour and infantry, while an attempted Israeli counter-attack from Hill 69 was driven back with serious loss. Elsewhere, the Egyptian Army's 1st Battalion took over the Palestinian villages of Falujah and Bayt Jibrin, while another detachment was sent east to link up with the Egyptian Light Force in the Hebron hills.

AIR OPERATIONS

The REAF's main focus on 2 June was in support of the Egyptian Army's effort against Negba, mainly this consisted of Spitfires bombing and strafing Israeli defensive positions. However, Egyptian Communiqué No.22, broadcast at 2200 hours GMT, also mentioned REAF attacks on the 'Zionist colonies' of Gan Yavne and Yibna, where hits were claimed on several buildings. In fact, Yibna was a Palestinian village whose population had largely fled to Isdud six days earlier. As the broadcast put it: 'Long columns of smoke were seen as the aircraft left their target. Our aircraft also raided Hulda on which high explosives and incendiaries were dropped, causing several fires … All our aircraft returned safely to their base'. Israeli sources added Rishon-Le-Zion and the major Israeli barracks at al-Sarafand to the list, raided in support of the Transjordanian Arab Legion.

Also on 2nd, or less likely 3 June, an unidentified aircraft flew high over Sde Dov airfield on the northern side of Tel Aviv, almost certainly taking a series of reconnaissance photographs. It was assumed to be Egyptian and was probably one of the REAF's Lysanders, a REAF reconnaissance photograph of Sde Dov taken on 2 or 3 June 1948 subsequently being published in an Egyptian magazine. Close inspection of this image indicates that it was taken with a camera having a 5 inch lens, at a height 18,000 feet,

An Israeli army convoy as seen underway in southern Palestine, in June 1948. (IDF)

and was number 198 in an otherwise unknown series. The sortie was almost certainly in preparation for an attempt to avenge the Israeli attack on Amman and although the reconnaissance aircraft was not intercepted, its presence may have forewarned the Israelis of an impending raid. Also published was an aerial photograph of the Israeli settlement at Negba, though without any additional information.

Meanwhile at Hilwan aerodrome south of Cairo, a new cohort of REAF Spitfire pilots were continuing their advanced training in No.6 Squadron. Amongst them was Pilot Officer Mustafa Wagdi Fadil who had probably been on leave or some other sort of training course since a 'local flying' mission on a Spitfire 5 (number 619) back on 11 May. Pilot Officer Mustafa Wagdi Fadil would subsequently be identified in the al-Arish Operational Diary as Mustafa Wagdi or simply as Wagdi. On 2 June he was back in the air, on a bombing-training sortie in a Spitfire 9 (number 642).

Here it is interesting to note that Spitfire number 642 would be mentioned in the al-Arish Operational Diary six days later, whereas No.6 Squadron would not be sent to al-Arish until 10 July. While the whole of No.6 Squadron was not rushed to the front in the aftermath of the Ramat David catastrophe, as previously thought, some of its aircraft and probably some of its pilots were. It is also likely that these urgent reinforcements, or more correctly replacements, included any members of No.2 Squadron who were not already at al-Arish.

3 JUNE 1948
GROUND OPERATIONS

At around midnight of 2-3 June, the Israeli Army's Givati Brigade, supported by an Irgun battalion and two companies from the Negev Brigade, attacked Egyptian positions north and east of Isdud. During this continuation of the Battle of Isdud, the attackers came under concentrated and accurate fire from alerted defenders. The attack ended in failure and the Israelis withdrew having lost, according to Egyptian estimates, about 400 killed and wounded. According to the Israeli soldier, historian and subsequently president, Chaim Herzog, this Israeli attack failed because: 'the plan was over-ambitious and co-ordination between the three attacking columns was faulty: the attack failed with heavy losses to the Israeli force including over 100 killed and wounded, and a number being taken prisoners'. The Egyptians lost 25 killed and wounded. Little comfort was provided by an Israeli radio statement on 3 June, which reported that: 'An Egyptian column, already attacked by the Jewish defence army, was bombed by Jewish aircraft last night. A direct hit was scored on a mechanised and armoured unit'.

Elsewhere, by the end of 2 June, Golani forces, which had been sent to counter an Iraqi threat to Lajjun, had captured further Arab villages on both sides of the al-Afula to Jinin road and in the Mount Gilboa range of hills. This opened the way for a Carmeli Brigade assault, which began on the night of 2-3 June, only to face stubborn resistance from Iraqi troops as Israeli units pressed into the hills

Mustafa Wagdi Fadil as seen while undergoing training at Almaza. Visible in the background are several North American T-6 Harvard advanced trainers of the REAF, including the serial number 414 (partially visible to the right), and a machine whose final number of the serial was 7, applied on the fuselage side, ahead of the roundel (visible to the left). (David Nicolle collection)

A Vickers Universal Carrier of the Egyptian Army (hull number 444) as seen during the advance into south-western Palestine, in May-June 1948. (Nour Bardai collection)

dominating the main road south of Jinin. Furthermore, the inability of the Alexandroni Brigade to launch its intended diversionary attack, coupled with the arrival of Iraqi reinforcements and the non-appearance of support from the Israeli Air Force, worsened the Israeli position further.

Having failed to take Negba, General al-Muwawi changed his priorities and instructed his officers to ensure full Egyptian control of a strip of territory from al-Majdal to Falujah and Bayt Jibrin before noon on 3 June. His main objective was to isolate Israeli forces in the Negev from their comrades farther north before the UN Truce came into effect. If successful, such a move would not only isolate the Negev but would enable Egyptian forces on the coast to reinforce their eastern colleagues holding territory from Beersheba through Hebron to Jerusalem. But, as General al-Muwawi had no available reserves, he had to divert his 1st battalion away from its attack on Negba to undertake this new task.

Other Egyptian units attempted to follow up earlier successes by pushing towards Bayt Daras and Be'er Tuvia, but here a sudden and uncharacteristic heavy summer rainstorm brought the ground assault to a squelching halt. The Egyptian Navy was also active on 3 June, shelling the port of Caesarea where a tug was claimed as having been sunk while other ships were also reportedly hit.

AIR OPERATIONS

As usual, the Israeli Air Force was more active at night than were its Arab opponents. According to the Voice of Israel at 0715 hours GMT the following day: 'Last night Jewish aircraft bombed enemy concentrations at Nablus and the police station was hit'. Thursday 3 June 1948 would be a day of starkly conflicting claims in the air-war between the Egyptians and Israelis, and it is now clear that the widely accepted version of events, based almost entirely on Israeli accounts, is fundamentally inaccurate.

The first REAF sorties on the morning of the 3rd were in support of the Arab Legion; a flight of Egyptian Spitfires attacking Israeli military headquarters, administrative centres and bases in the area of al-Sarafand and Rishon-Le-Zion south-east of Tel Aviv. Sarafand al-Amar, or Sarafand al-Kubra, used to be the site of the British Army's largest base in the Arab Middle East, but was taken over by Israeli forces around 19 or 20 May 1948. According to Egyptian Communiqué No.23, broadcast from Cairo at the usual time of 2200 hours: 'Our Air Force concentrated its raids on the fortified colony of Rishon-Le-Zion where our heavy and light bombers, with a strong fighter escort, dropped high explosives and incendiaries, demolishing several buildings. Our airmen, during their successive raids against this colony, observed big fires which began to spread'.

Egyptian sources make no mention of an intervention by Israeli fighters, and the al-Arish Operational Diary for this day is not yet available. Israeli sources, however, state that Modi Alon, the leader of the IDF/AF's only fighter squadron, No.101, took off from Aqir in an Avia S-199 (number D-120) and engaged one Spitfire. He claimed to have 'poured machine-gun fire into it' over Beit Hanan (Beit Hanan just south of Rishon-Le-Zion, rather than Bayt Hanun north-east of Gaza) before the Egyptian pilot escaped.

Another REAF mission that morning, shortly before or slightly after the attacks on al-Sarafand and Rishon-Le-Zion, targeted Israeli positions in Gan Yavne east of Isdud, along with Qastina 38km north-east of Gaza. These were in close support of the Egyptian Army, Communiqué No.23's account stating that: 'Our Air Force also strafed an enemy convoy near the colony of Gan Yavne, destroying six armoured cars, a large number of lorries and passenger cars as well as a large concentration of vehicles north-east of Qastina'.

The results of the REAF's bombing raid against Tel Aviv in the late afternoon or early evening are much more controversial. Furthermore, it is unclear whether the attack was in response to King Abdullah's request for retaliation, which seems had already been carried out, were a continuation of it, or were simply strategic in intention. Two official Egyptian Communiqués mentioned the raid, No.23 broadcast at 2200 hours GMT on the 3rd stating that: 'Another raid was carried out against Tel Aviv where bombs were dropped on the airfield and main electricity power station, which received direct hits. All our aircraft, except one, returned to their bases'.

An Egyptian reconnaissance photograph of Sde Dov airfield, outside Tel Aviv, taken on 2 or 3 June 1948. (David Nicolle collection)

A still from an Egyptian newsreel ('With the Egyptian Army in Palestine'), released following the capture of the Kibbutz Yad-Mordechai, on 24 May 1948, showing three of the REAF C-47 Dakotas in formation. In combat, the Egyptians never deployed as many Dakotas in a single raid, nor had they flown in such relatively large formations. Therefore, this formation must have been staged for the benefit of the cameraman in the fourth C-47, probably during a ceremonial flypast. (David Nicolle collection)

The following day, at 2245 hours GMT, Egyptian Communiqué No.24 updated the previous report, stating that: 'The aircraft which was reported yesterday unable to return to base after completing its mission, had engine trouble but the pilots landed it safely at our front lines, all the crew surviving'.

Before looking at other conflicting accounts of the fate of two REAF C-47 Dakota 'bombers' involved in this raid, a summary of what is now known to have happened is helpful. After the departure of the UN Mediator from Tel Aviv, the Egyptians certainly sent two Dakotas against that city. They carried their usual external bomb-load, but were otherwise unarmed and reportedly lacked a fighter escort (contradicting Israeli sources) because they hoped to approach and escape under the day's extensive cloud cover. As they approached Tel Aviv, the clouds dispersed and they became exposed to any Israeli fighters that could reach them. The crews released their bombs on the assigned targets, but before they could leave the area they were intercepted by an Israeli Avia S.199, flown by Modi Alon. Both Egyptian aircraft tried to escape out to sea but one was hit and subsequently made a forced landing near Isdud.

The triumphalist tone and inaccurate facts of Israeli accounts of this action became apparent soon after the event, the Voice of Israeli radio station announcing that: 'An enemy bomber was brought down by a Jewish fighter aircraft north of Tel Aviv at 1930 today; it was blasted in mid-air and pieces fell near Jaffa. An enemy supply plane was brought down over Rehovot at 1900 by a Jewish fighter aircraft'. The circumstances of the earlier claim remain unknown, but Rehovot was south of Rishon-Le-Zion. If the BBC Monitoring Service mis-heard or mis-transcribed the time mentioned, it might be a reference to Modi Alon's earlier clash with an Egyptian fighter over Beit Hanan.

Later Israeli accounts of Modi Alon's interception of REAF Dakota 'bombers' near Tel Aviv offer further colourful details, some being exaggerations of reality while others were simply untrue. At an unknown date the *Palestine Post* newspaper also stated that one of the Dakotas supposedly shot down by Modi Alon 'blew up' in mid-air, with fragments falling on Jaffa. The often-unreliable Leo Nomis, who himself flew Israeli Avia S-199 fighters, based his account upon that of Alon's himself. He wrote that Modi intercepted the two REAF Dakotas after they had turned for home, that the Egyptians broke formation when attacked and took different courses.

Undeterred, Modi Alon climbed above them before diving to attack the first Dakota. According to Nomis, the second Dakota crossed the coastline, Alon closed in behind and opened fire. Flames appeared beneath one wing and expanded until they enveloped the entire wing. Meanwhile down below, thousands in the streets of Tel Aviv noted how bright the flames were against the blue sky. 'The doomed machine rolled down towards the calm surface of the sea and pieces began to fall from the fiery wreckage. A plume of water and spray and black smoke briefly marked the place where is disappeared'. As a result, Modi (Mordecai) Alon was credited with the destruction of two Egyptian C-47 Dakotas while flying an Avia S-199.

Today, the currently accepted version of the interception has Alon easily avoiding either two or four REAF Spitfires, then shooting down one Dakota in the first air combat victory claimed by the Israeli Air Force, and probably downing the other. The destroyed Dakota supposedly crashed on the beach a few kilometres south of Bat Yam, itself just south of Tel Aviv, and the crew were either killed or baled out before the crash, while the second Dakota came down within Arab-held territory at the mouth of the Wadi Sukhrir between Isdud and Yibna, the crew surviving the crash. Thereafter there was supposedly little further Egyptian air activity before the First UN Truce came into effect, with no more raids on Tel Aviv by REAF Dakotas.

Rather peculiarly, considering the importance of such a feat, no evidence – like pieces of wreckage – was ever presented by the Israelis. Arguably, as some sources have suggested, the second Dakota was recovered by the Egyptians. While, the official Israeli version and the subsequently accepted one contain major errors, their conclusions were not surprising given what was reported by Modi Alon and what was seen from the ground. Abd al-Rahman Inan, the captured REAF pilot at Camp Yona, also heard the combat and similarly believed that both the Dakotas had been downed. Not until he was released at the end of the Palestine War and returned to Cairo in 1949 did he learn the truth. As Inan later recalled:

> One afternoon as I was sitting on the sand near a shelter, I hear our aircraft and saw two aeroplanes coming from the south, but the siren remained silent in spite of the sound of aircraft becoming close. The soldier look(ed) at me… He said, "Do

you hear (the) sound of aircraft, I think you recognise your own aircraft", and before he finished his question bombs start falling. The soldiers all went into shelters and I watched for the aircraft eagerly and my heart danced for joy when I saw them. And when this ignoble soldier saw me happy, he pushed me into the jail. When I entered the room, I could hear the sound of a small aircraft flying at low level. Then I hear clapping and cheering from the soldiers, but their voices were mixed because they were all speaking at the same time. Then the ignoble soldier came and told me that a Messerschmitt had attacked two Egyptian aircraft and shot them down into the sea. I did not sleep that night wondering whether it was true or false … I did not learn the truth until I returned to Egypt and I found that both aircraft returned safely to their air base (in fact both crews returned but not both aircraft) and not one of our bomber pilots was hurt (in combat) throughout the Palestine War.

On 8 September 1948 the Egyptian newspaper *Akhar Sa'a* published a different version of the event. One of its best-known reporters, Muhammad al-Bayaly, had interviewed a then unnamed REAF Squadron Leader, now believed to be Muhammad Sidqi Mahmud al-Milaigy, who told him the following:

We were attacking Tel Aviv when Jewish anti-aircraft fire suddenly opened up. We were at a low altitude, and I felt the bottom of my plane on fire. Three of my crew were bleeding. I descended at a crazy speed having made several direct hits in the port area. The smoke made it almost impossible to breath and the flames were devouring the rear of the plane. My only concern was to reach the nearest Egyptian location as quickly as possible before the Jewish gangs caught us.

Once we got down and left the plane, its fuel tank exploded. Minutes later one of the enemy aeroplanes flew over us (presumably Modi Alon) so we threw ourselves onto the sand. This plane flew over several times then returned home. Jaffa and Tel Aviv were in sight and we were exhausted. Nevertheless, with help from an Arab tribe (or family?) which had been driven from inside Palestine, we escaped… A little boy named Hassan led us to them and swore on the Koran to offer us safety. We asked then where we were and were told that we were only 6 km from Isdud.

The burning C-47 Dakota had crash-landed in the Arab Sukhrair, a broad area of coastal sands, probably close to a narrow strip of firm ground along the lowest reach of the Wadi Sukhrir. This normally dry stream currently formed a notional front-line between Israeli-held and Egyptian-held territory.

On 1 June 1949, the head of the REAF published another article in the REAF's own magazine. This indicated that the Dakota's navigator was the most seriously injured, that the machine's left wing caught fire, burning the aeroplane's tail and that the flames spread farther before the pilot could make his emergency landing. Muhammad Sidqi Mahmud al-Milaigy was interviewed about this incident in the late 1950s, by which time he was at the head of what had been renamed the Egyptian Air Force. When asked what had been the most difficult moment during his time as a pilot, Sidqi Mahmud recalled the raid on Tel Aviv on 3 June 1948, stating that:

The most embarrassing moment was when I had to take the decision to make a force landing in enemy territory … I remember the details very well … My aircraft was on fire and had some wounded crew in it … I had two options, … either to risk my aircraft and continue flying until I reached the base … which would avoid being captured by the enemy. But doing this would mean gambling with the wounded men's lives. Or I could make an emergency landing in (what he thought was) enemy territory … Within seconds I decided on an emergency landing … We descended and by the will of God I landed with the aircraft on fire. With the wounded we had to walk to the base (in reality only as far as the closest Egyptian Army position) and succeeded in reaching it without being captured.

Camp Yona, on the northern side of Tel Aviv, where Haganah trained its new recruits – but also kept captured REAF pilots. (Photo by Beno Rothenberg, via Israeli State Archives)

The RIrAF made a big effort on Thursday 3 June to support Iraqi troops who were now under considerable pressure from an Israeli offensive. These were, however, not the same sorties as the RIrAF south of the Dead Sea in support of Transjordanian troops. The latter were reported by the Arab News Agency at 1745 hours GMT on the 3rd: 'An Amman report stated that Egyptian and Iraqi aircraft bombed and fired on Jewish concentration(s) in Ghur al-Safi, on the eastern bank (in fact immediately south) of the Dead Sea, where the police station was re-occupied, the Jews suffering heavy losses during their evictions'. Israeli sources described this attack as being carried out by Iraqi Ansons that targeted Sdom (Sodom) from high altitude and made no mention of Egyptian participation.

The Syrian Air Force was also back in action around Ashdot Ya'akov south of Lake Tiberius on 3 June, an official Syrian communiqué reporting that: 'Our Air Force carried out extensive reconnaissance duties in various parts of this sector'. According to Egyptian Communiqué No.23, broadcast at 2200 hours, during an Israeli attack in the Negba area: 'Our anti-aircraft guns fired at a Zionist plane which was cooperating with the enemy during this battle and brought it down in flames'. Like the Israeli claims over Tel Aviv, this Egyptian claim appears to have been mistaken.

4 JUNE 1948
GROUND OPERATIONS

Early on 4 June, the Iraqis launched a successful counter-attack against the Carmeli 1st Battalion in the hills southwest of Jinin and, as all efforts to relieve this unit failed, the Israelis were forced to withdraw. Indeed, by the evening of the 4th, the Carmeli situation would become desperate. On the southern front, however, Israeli forces took control of the abandoned Palestinian village of Yibna.

AIR AND NAVAL OPERATIONS

Earlier on Friday 4 June, at 0230 hours GMT, Cairo Radio broadcast a joint Arab statement concerning foreign aircraft flying over Palestine. This announced that:

> 'According to the Amman correspondent of al-Ahram (newspaper), it had been proved that the Zionist were using foreign aircraft in their raids. As a result of contacts between the military command and competent authorities in the Arab countries, a warning has been issued that any foreign aircraft flying over the combat area in Palestine will be considered an enemy aircraft and is liable to be fired on'.

Paradoxically, it now seems that within a few hours the REAF would itself send an aeroplane with foreign markings into combat.

According to Air Commodore Hayes, the British Air Attaché in Cairo, in his first Annual Report on the REAF (covering the period from 1 January 1948 to 1 January 1949), the British arms embargo also came into 'total effect' on 4 June 1948. Of perhaps more immediate concern to Egyptian service personnel at the time was a clash between the Egyptian Navy and the Israeli Air Force, which also drew in the REAF. Reportedly, at 1415 hours Cairo time, a naval convoy consisting of the Royal Egyptian Navy's armed transport *Malikah Fawzia* (also called *Al-Amira Fawzia*) with two 3-pounder guns, and perhaps another unarmed transport and a landing craft, was seen by Israeli observers heading north along the Palestinian coast. An attempt by the Israeli corvette *Eilat* to stop them failed, though she reportedly exchanged fire with the *Malikah Fawziya*.

At Sde Dov airfield, three aircraft of the Israeli Air Force, a Beech V-35 Bonanza, DH-89A Dragon Rapide and what was reportedly a Fairchild F-24R Argus, were meanwhile loaded with bombs, took off and headed towards the Egyptian ships. The REAF at al-Arish was also ordered to support the *Malikah Fawziya* and her charges. The events that followed remain confusing, but clearly resulted in the loss of the Israeli Argus. This machine is said to have previously had the Egyptian civil registration SU-ADR before it was seized by a Zionist paramilitary unit, having been ordered to land by the British after overflying a restricted area in Palestine (to be described in detail in the Volume 12). But SU-ADR was an RWD-13. Israeli accounts maintain that the Argus (or maybe RWD-13) was hit by anti-aircraft fire during its second bombing run over the Egyptian ships, crashing into the sea at 1730 hours Tel Aviv time with the death of its crew, David Shprinzak and Matetiahu Sukenik. Nevertheless, the Egyptian ships withdrew to the south. Meanwhile, another Israeli airman, Boris Senior, in the Bonanza, was convinced that at least one of his bombs had struck an Egyptian ship.

The Egyptian version, as broadcast at 2245 hours GMT, presented a very different story, which was not only misleading but contained several errors. Thus, official Communiqué No.24 stated that: 'The Royal (Egyptian) Navy has continued the operations started yesterday (when there had been a reported naval attack on Caesarea harbour). It was attacked by three enemy Messerschmitt-type planes, so our fighters took off, shot one down and routed the others'. In reality, when the Egyptian convoy was some distance from Jaffa, it was attacked by Israeli aircraft which made six passes or bombing runs but failed to make any hits. Meanwhile those Egyptian ships that had anti-aircraft guns fired back. The Egyptian naval commander must have already requested air support when challenged by the *Eilat*, because a REAF fighter arrived on the scene while the Israelis were attempting to bomb the Egyptians.

Squadron Leader Muhammad Abd al-Hamid Abu Zaid, the pilot of this Egyptian aircraft, then reportedly shot down the Argus or RWD-13 while the other two fled. It was his first and perhaps only confirmed air combat victory, though he was eventually credited with one or more "probables", plus those that he certainly destroyed on the ground. Abu Zaid may also have been flying the Hawker Fury prototype that had been commandeered by the Egyptian authorities on 24 April 1948. It still had its British civil registration G-AKRY when was sent to the REAF's front-line aerodrome at al-Arish at an unknown date, though it did not officially join the REAF until late in June. Only Squadron Leader Abu Zaid was allowed to fly this potent and now armed machine on operational flights, although Tahir Zaki may have had collected some experience at its controls, too. If Abu Zaid did indeed fly the Fury to the aid of the Egyptian Navy, he would have done so when it was still in British civilian markings.

At 1830 hours GMT, the Voice of Israel broadcast the first official Israeli version of events: 'A naval battle was fought off the coast of Tel Aviv today … Planes of the Israeli Air Force were sent to bomb the enemy warships and at least one direct hit was scored. After the bombing attack, the enemy ships turned and sailed away. The battle lasted about three hours. One of our planes has not returned'. Hoping that the downed airmen might have survived, Boris Senior flew three further sorties in the Bonanza, searching for his colleagues. On one such flight he was joined by Adam Shatkai in a Dragon Rapide, but without success.

On 29 December 1976, the author received a letter from Colonel Y. Carmali, the Israeli Deputy Defence Attaché in London, which provided a revised Israeli assessment of the engagement. Colonel Carmali stated that:

> During the first week of June '48, the IDAF lost one aircraft of Fairchild type. Three of our airplanes were scrambled to intercept Egyptian vessels that tried, according to our estimate, to land forces at Ashdod (Isdud) or to shell Tel Aviv. The airplanes carried out two runs on the ships. During the second run, the Fairchild was hit by AA fire and hit the sea at 1730 hrs. The pilot David Shpinzak and the bomber Matetiahu Sukenik were killed.

More recently released Egyptian documents now show that this naval operation had indeed been ambitious. Around midday on 4 June a task force of three vessels – consisting of a minesweeper (perhaps the *Malikah Fawzia* had been re-equipped for this role or there had been some renaming of Egyptian naval vessels), a landing craft and a heavy transport boat – moved towards Tel Aviv. Its mission was to destroy harbour installations and facilities, and to mine the entrance. This Egyptian task force was observed by the Israelis as it moved northward along the southern part of the Palestinian coast, so the *Eilat* put to sea while air support was prepared. The Egyptian task force and the *Eilat* sighted each other about 20km north of Gaza and 7km from the coast off al-Majdal.

As its air support had not appeared, the captain of the *Eilat* either decided, or was ordered, not to tackle this Egyptian force alone. In fact, at 1400 hours the *Eilat* was instructed to return north, but as soon as the Israeli ship did so the Egyptian minesweeper increased speed, and soon closed to less than 5km and opened fire. The *Eilat's* stern was hit though the damage was not severe. When the Israeli ship returned fire, its shells exploded at half-range. Thereafter the two vessels exchanged fire without any further effect for about one hour. Even so, with the element of surprise lost, the Egyptian ships were told to return home.

For the rest of the Palestine War, the Egyptian Navy, though occasionally bombarding Israeli coastal installations, was mainly engaged in providing logistical support to the Egyptian Expeditionary Force. Meanwhile, Israeli naval activity was confined to patrolling the approaches of Tel Aviv harbour and escorting in-bound immigrant ships.

Muhammad Abu Zaid's victory had meanwhile been seized upon by the Egyptian media in an effort to maintain public support for the campaign in Palestine. It would also feature prominently in an article about 'Death's Head Squadron Leader Abu Zaid', written after his loss later in the war by an upcoming Egyptian journalist named Muhammad Hassanayn Haykal – later to be the mouthpiece of President Nasser's governments in the 1960s.

According the Haikal's colourful, and not entirely reliable account, one of the Egyptian ships sent an SOS directly to al-Arish aerodrome:

> Unfortunately, a sand storm had already blown up, and this usually grounded all the aircraft. No pilot was ordered to go and help the Egyptian ship. Instead, the squadron was called to a quick meeting and a volunteer was asked for. Sq. Ldr. Abu Zaid immediately stood up and said that since he was the one who got the message, he should be the one to go. In fact, he then left the ready room before anyone else could speak. He took off in the middle of the sand storm and headed straight for the coast. It was 3 o'clock in the afternoon. The Egyptian ship was very exposed and was under attack from the Israeli aircraft. One of the Israeli aircraft was circling the ship when suddenly they (the Egyptian crew) saw an Egyptian aircraft approaching. It was Muhammad Abd al-Hamid Abu Zaid. All the sailors on the ship recognized his aircraft and so they started to shoot at the Israeli. This (Israeli) aircraft was claimed shot down and then Abu Zaid disappeared from view again …
>
> Two Israeli planes were then flying around this area, uncertain what to do. Abu Zaid started to shoot at one of them, which reportedly caught fire while the second fled back to Tel Aviv. Abu Zaid then disappeared for the third time and was thought to have headed back to his base. The sun was bright over-head and Abu Zaid did a victory roll in salute. The ship blew its whistle three times to thank him for his help. Abu Zaid flew back towards his base, facing the sand storm again.

Elsewhere during the afternoon of 4 June, REAF Spitfires were sent to support Palestinian militia irregulars who were blockading the Israeli settlement and airfield of Nir Am. Here one low-flying Spitfire was hit by ground-fire and reportedly force-landed near al-Majdal, although the pilot escaped to rejoin his squadron. A peculiarity of this account is that al-Majdal is 10 km due north of Nir Am, in almost the opposite direction a damaged Egyptian aircraft would fly if its pilot wanted to get home safely.

When the British government learned that the Hawker Fury prototype had entered combat while bearing British civilian markings, they were understandably furious. The Egyptian Prime Minister was told of this by the British ambassador, agreed that it was unacceptable and promised to investigate the matter. The Egyptians therefore gave the machine proper REAF markings and the serial number 701. The red fuselage stripe was removed and the front part of the red spinner was repainted black with a white skull-and-cross bones, like that of Abu Zaid's earlier Spitfire. Nevertheless, this repainting would not be done until after the First UN Truce had come into effect.

At the end of an eventful day, Egyptian Communiqué No.24, at 2245 hours GMT, merely announced that: 'Air Force activity was restricted today by bad weather conditions and a sandstorm in the operational area. From all these operations, all our aircraft returned safely to their bases'. This directly contradicted the Israeli claim that the Spitfire that made a forced landing near al-Majdal came down on 4 June.

Elsewhere, both the RIrAF and Syrian Arab Air Force (SyAAF) were in action. According to an Iraqi communiqué concerning operations on 4 June, broadcast by the Arab New Agency at 1000 hours GMT: 'Our Air Force raided the eastern industrial district of Tel Aviv, there great explosions were observed. Direct hits were scored on the potash plant south of the Dead Sea, and enemy concentrations were displaced'. The latter mission suggests that the attack on Sdom the previous day had either been repeated, or this communiqué was repeating the previous day's target. At 0900 hours GMT on 5 June, the Arab News Agency was almost certainly referring to events on the 4th when it reported that:

> Our (Iraqi) Air Force attacked an enemy mobile column north of the Jinin area, destroying most of the vehicles. Tel Aviv's industrial suburb of Petah Tikva, on which the Jews largely depend for the production of their military equipment and ammunition, was heavily bombed with incendiary and high-explosive bombs. Many buildings were destroyed and the fires could be seen from a great distance, burning for hours. All our aircraft returned safely to their base.

The REAF and the RIrAF were thus now operating in the same area, attacking the industrial zone east of Tel Aviv. This might have been

Pilots and ground crew preparing for a training sortie in the front of a pair of Spitfire Vb/c of the REAF. Notable is the aircraft in foreground, and the black front half of its spinner: this was probably the Spitfire on which Abu Zaid applied the skull & bones insignia. (Dr David Nicolle Collection)

why black and white stripes, comparable to but smaller than the 'invasion stripes' painted on Allied aircraft in the latter part of the Second World War, would soon be added to most Egyptian and some Iraqi combat aircraft. There is, however, no evidence of friendly fire incidents involving these two Arab air forces, both of which were up against a rapidly strengthening Israeli Air Force which used similar and, in the case of the Egyptians, soon identical aircraft.

Farther north, Iraqi Ansons attacked Israeli targets in the Netanya and perhaps the Afulah areas on the 4th, having also possibly attacked al-Afulah earlier in the day. However, according to Syrian Communiqué No 20, broadcast from Damascus at 2015 hours. 'Our Air Force bombed the settlement of al-Afulah, destroyed enemy instillations and started fires. Our Air Force also attacked a group of cars south-east of the settlement, destroying a large number of them'. Farther to the north, SyAAF Harvards struck more habitual targets, hitting Israeli military positions at Rosh Pina and al-Malkiyah on 4 June, while a Palestinian broadcast from Jerusalem at 1300 hours reported that: 'A Jewish aircraft flew over Nablus at about 0400 this morning and dropped about 30 bombs, all of which fell in the streets and open places of the town, wounding one civilian'.

5 JUNE 1948

GROUND OPERATIONS

Before midnight on the night of 4-5 June, Colonel Moshe Carmel, the commander of the Israeli northern front, advised Haganah High Command that 'holding on in this desperate situation would be worthwhile only on the condition that a major effort was made by the Alexandroni Brigade to capture Tulkarm'. Due to heavy Israeli losses and the inability of the Alexandroni Brigade to accomplish its mission, Colonel Carmel was given permission to withdraw. This Israeli forces did under Iraqi fire, leaving Jinin before dawn on the 5th but retaining their positions north of the town. Meanwhile, on the Lebanese Front, the Lebanese attack struck al-Malkiyah from the east and the west simultaneously while the ALA battalion pressed Israeli Oded troops from the south. In fact, the Israelis were caught by surprise and after some delay in front of the minefields the Lebanese force broke through Oded's positions.

AIR OPERATIONS

Saturday 5 June would see less activity in the air than the 4th, though the Egyptian communiqué broadcast from Cairo at 2315 hours GMT reported that:

> Our Air Force continued its armed reconnaissance flights, obtaining valuable information on enemy concentrations and movements. The settlements of Yavne and Gan Yavne were bombed with high explosive and incendiary bombs, big fires being started. As a result of Air Force reconnaissance flights over Khulda, Hulda and Abu Shusha, our aircraft dropped high explosives and incendiaries on enemy concentrations and machine-gunned a convoy, damaging many of the vehicles. All our aircraft returned safely to their bases.

At 1825 hours, the Voice of Israeli merely stated that: 'Enemy aircraft bombed the Givat Haim area'. This lay south of Hadera on the coastal plan between Tel Aviv and Haifa, and the aircraft involved may have been Iraqi. RIrAF Ansons certainly bombed an Israeli concentration in al-Afulah. Meanwhile Syrian Harvards focused their attentions on eastern Galilee and according to Syrian Communiqué No.21, broadcast at 2050 hours:

> Our Air Force bombed the settlements of Najmat al-Subh (the land upon which Ayelet HaShahar was constructed), Rosh Pina and al-Malkiyah, on each of which direct hits

> were scored, inflicting heavy losses and damaging depots and supply centres. A squadron of our Air Force attacked enemy lines north of al-Malkiyah, causing panic among the enemy. Other Air Force squadrons attacked enemy camps east of al-Malkiyah and dive-bombed a group of transport cars, destroying a large number of them. The Air Force also supported our troops in their advance towards al-Malkiyah. Our aircraft bombed the settlement of al-Manara, hitting several enemy positions and damaging water tanks. The settlement of Rosh Pina was bombed again and fired (sic.) were started. The settlement of Kfar Giladi was also bombed. Tuba (perhaps Tuba al--Zanghariyah where the Arab bedouin inhabitants supported the Haganah), Ein Gev, Najmat al-Subh (Ayelet HaShahar) and the airfield of Rosh Pina were bombed by another squadron. Our Air Force also attacked enemy concentrations in these centres, inflicting considerable losses in men and equipment. Ammunition depots were set on fire and explosions followed.

The number of targets attacked was large, though Israeli sources show that Syrian claims of damage inflicted were inflated. The communiqué's references to several squadrons being in action is also very misleading. There were only two SyAAF Harvard units, and the sorties listed in Syrian Communiqué No.21 mostly involved no more than two aircraft. Even so, this had been an impressive effort by the SyAAF.

6 JUNE 1948
GROUND OPERATIONS

Having occupied the lateral axis, Egyptian General al-Muwawi focused upon securing his flanks and lines of communication. The first objective was to be the settlement of Nizzanim, a formidable obstacle about midway between al-Majdal and Isdud, which had been bypassed during the initial advance. It was garrisoned by a force of 150 local militia, reinforced by a Givati force and supported by Israeli artillery from Be'er Tuvia. The topographical features of Nizzanim consisted of three elevations, each with a ring of fire position and communication trenches. In addition, the settlement was surrounded by sixteen fortified fire positions, eight of which were equipped with heavy machine guns while the others were equipped with anti-tank weapons. Beyond these, the outer ring of the settlement had four belts of barbed wire and minefields.

During the night of 6-7 June, Arab irregulars harassed the defenders of what was in reality a fortress, in preparation for the main attack. This was assigned to the Egyptian Army's 9th Battalion supported by a light tank battalion (13 tanks), an armoured car company and two artillery batteries (each of six 25-pounders). Learning from their failures and their successes during previous battles, the Egyptian assault was carefully planned and would be carried out in two phases. The first aimed to capture a hill north-east of the settlement, on which water reservoirs had been constructed, and then an advance to capture the north-eastern sector of the main settlement. The second phase aimed to capture the rest of Nizzanim.

On the northern front, while the Lebanese were attacking al-Malkiyah early in the morning of 6 June, the Syrian Army moved two infantry battalions against Mishmar HaYarden just opposite the Jisr Banat Ya'qub ('Bridge of Jacob's Daughters'). Their objective was to cut the north-south road that connected the Finger of Galilee with the centre of Galilee, then to join ALA and the Lebanese units near al-Malkiyah. However, due to accurate Israeli mortar and artillery fire, which had pre-targeted the river fords in front of this settlement, Syrian armour failed to get across the river Jordan. So, the two infantry battalions, which were now harassing the settlement, also withdrew because the Syrians needed to reorganise themselves.

Meanwhile the Israeli Oded Brigade was largely concentrated in the vicinity of Mishmar HaYarden east of the river Jordan. Therefore, while the battle was under way, two further Syrian contingents mounted diversionary attacks to divert the Oded and Golani Brigades away from the main Syrian thrust. The first was on 6 June, against the Dan settlement at the northern tip of the finger of Galilee. Meanwhile, part of al-Malkiyah fell into Lebanese hands for the second time by the evening of 6 June. According to a Lebanese communiqué broadcast from Beirut at 2145 hours: 'The number of (enemy) killed in the two-day al-Malkiyah battle is estimated at 132, and the number of wounded is probably many times that number, as the Syrian Air Force dealt very effective blows (in support of the Lebanese Army)'. According to a Voice of Israel broadcast three hours earlier: 'Despite the large number of attackers, supported by heavy artillery and aircraft which bombed this village for a long time, the enemy failed to eject the Jewish forces from any position in the village except the customs house, which was evacuated'.

Egyptian Army field artillery (probably a 25-pounder) as seen in action in southern Palestine in 1948. (David Nicolle collection)

AIR OPERATIONS

At the end of Sunday 6 June 1948, Egyptian Communiqué No.26 broadcast at 2200 hours GMT merely stated that: 'The Air Force made reconnaissance flights beyond our forward positions, but adverse weather and sandstorms limited its activities. All aircraft returned safely'. According to an early account of Modi Alon's life, three days after he was credited with shooting down the Egyptian Dakota or Dakotas – therefore on 6 June – he attacked some REAF Spitfires and downed one, but was shot up by another. Alon then supposedly landed back at his home aerodrome slightly injured. Subsequently, this version of events was vehemently denied by all Israeli sources, but if it actually took place two days later, on 8 June, it would sound more credible (see below).

No mention seems to have been made of Iraqi air activity on this day, but the SyAAF was again in action. According to Syrian Communiqué No.18: 'A squadron of our Air Force continued its raids against Nahariya where a considerable part of the warehouses and quays were damaged and two vessels were sunk. Our squadron left their target with columns of flames rising from it. Another squadron attacked the Jewish colonies of Dan and Kfar Giladi, where the targets were heavily bombed'.

Sunday 6 June was also the date of the first entry in SyAAF pilot Faisal Nassif's diary, found on his body when he was shot down later in the war. Nassif was currently still under training at al-Mizzah aerodrome on the southern outskirts of Damascus. This highly personal document also sheds a touching light upon the life of a newly married young man who was already facing tension with his in-laws: 'We had a quarrel with Alice's (his wife's) parents and left the house al-Mizzah. I took her to my parents' house. I went with my mother to buy clothes for her'.

7 JUNE 1948

Count Bernadotte, the official UN Mediator, who had earlier flown to Amman to speak to King Abdullah, returned to Cairo the same day where he had further discussions with the Egyptian government. Then he travelled back to Amman, on to Haifa, to Cairo once more, and finally to Beirut. These exhausting talks with Arab and Israeli government representatives were, however, making progress. The following day, 7 June, the entire round-trip was repeated, this time flying to Beirut, Haifa, Mafraq, Amman, Mafraq a second time in one day, Haifa once more, and back to Cairo. While there had been no flight to Iraqi territory, the Iraqi military were based at Mafraq, which Bernadotte visited twice. Now, at last, all sides agreed to a ceasefire, and it would come into effect at midnight on 11 June. Before then, however, there would be more fighting on the ground and in the air – some of it intensive.

GROUND OPERATIONS

On 7 June, the Egyptian Army launched a determined attack north of Isdud and, supported by artillery and the REAF, forced the Israeli front-line back to a line of abandoned Arab villages that had already been fortified. While the Egyptian forces stopped to regroup, they continued bombarding the Israelis with artillery and from the air. The Egyptian Army also achieved success in its assault upon the Israeli settlement of Nizzanim. According to Egyptian Communiqué No.27: 'The attack, made from all directions, and supported by artillery and the Air Force, was completely successful. The enemy collapsed, the settlement was occupied and 110 prisoners-of-war were taken'.

The attack had started with a raid by at least two REAF Spitfires, which bombed and strafed Nizzanim's defences at 0900 hours, accompanied by an artillery bombardment. An hour later Egyptian light tanks advanced to suppress the settlement's fortified fire positions, while two infantry companies infiltrated the ruins of a British army camp north-east of the settlement. As soon as the fortified fire positions were suppressed, and the engineers had opened breaches through the minefields and barbed wire, the two infantry companies attempted to break into the settlement but they faced stubborn resistance. Nevertheless, after severe fighting, the Egyptians succeeded in penetrating the Israeli defensive positions and by noon they controlled the north-eastern part of the Nizzanim.

To exploit this success, Colonel Kamal al-Rahmani, the Commander of the 9th Battalion, committed the rest of his force to capturing the rest of the settlement. Israeli artillery from Be'er Tuvia shelled this attacking force, but could not prevent it from pushing deep inside the settlement, gradually driving the defenders southward. When the latter realized that they could no longer hold Nizzanim, they attempted to withdraw through an orange grove but found themselves hemmed in by Egyptian forces. At 1630 hours, the settlement surrendered having suffered 33 killed, plus 110 now taken prisoner. The rest of the garrison escaped. For their part, the Egyptians suffered 13 killed and 45 wounded.

Next day the Voice of Israel admitted this loss in its news bulletin broadcast at 1730 hours GMT: 'On the southern front, Egyptian forces supported by planes, tanks and heavy artillery have for several days been attacking the small and isolated Jewish settlement of Nizzanim which the enemy succeeded in capturing yesterday'.

The Douglas C-47 Dakota of the UN, frequently used to fly Count Bernadotte, as seen in Palestine in 1948. (Albert Grandolini Collection)

AIR OPERATIONS

Having suffered losses to ground-fire which, though few in number were significant for the small REAF Tactical Air Force, the Egyptian Spitfire pilots had been ordered to carry out their ground support missions from a greater height. This caused frustration, not least amongst young men who were still early in their REAF careers. One such was *Tayyar Awal* Abd al-Majid al-Rafa'i. The following eventful sortie, which he later recounted to Munira Kafafi, who worked for Egyptian television and was the daughter of a bomber pilot, probably took place during the attack on Nizzanim. Here it is important to note that REAF promotions up to rank of Squadron Leader (like those up to a comparable rank in the Egyptian Army) were awarded annually in June, before being made public in July.

As Munira Kafafi wrote:

> The Egyptian aircraft became renowned at that time for dexterity at low altitude. Obviously, this sort of work made the pilots and aircraft more exposed to danger, so the pilots started flying even lower though this exposed them to the danger of hitting hills. Clearly the Egyptian leadership recognised the importance of this danger and so sent orders stopping the men from flying so low. But Lt. Gen Abd al-Majid al-Rafa'i was deaf to these instructions. He smiled when he related to me the following story. 'Naturally my colleagues and I stopped, and we then flew at higher altitude. On the day I was promoted to Tayyar Awal (First Pilot or Flight Officer) I was sent to attack one of the settlements. The Egyptian aircraft were unlucky when hitting the position and reported that it was unoccupied by troops, but a doubt entered my soul and I descended, despite the warnings of our leadership, and my aircraft nearly hit a haystack. But do you know what I found? All the garrison of the settlement had gone into shelters which were hidden between sheafs of corn. So, our Egyptian aircraft returned to demolish the place entirely and thus complete the task assigned to us'.

The more experienced REAF Spitfire pilot Tahir Zaki similarly recalled the siege and fall of Nizzanim: 'Another kibbutz was called Nizzanim and it was closer to al-Majdal. I asked General Muwawi to let me attack its water tower with bombs again. I thought this would reduce Egyptian casualties in their next attack. In fact, our Army then took the place with very few losses'.

The taking of Nizzanim would feature prominently in Major Muhammad Kamal Siddiq analysis of the REAF's role in the first part of the Palestine War. Entitled 'Air War in Palestine', it was published in the October 1948 edition of the *REAF Magazine*, towards the end of the Second UN Truce. Here Major Siddiq wrote that: 'In the Battle of Nizzanim (6-8 June), the Air Force also bombed it, based on a tight joint-plan developed by the headquarters. In this battle, the air force was used as a shield to cover our infantry (ie, as a protective curtain), resulting in low casualties amongst the ground forces'.

An article by Lieutenant Ali Muhammad Mahbub in the same edition of the *REAF Magazine* seems to have been based upon the experiences of those involved, perhaps including Mahbub himself. He wrote that reconnaissance confirmed that the garrison of Nizzanim had been reinforced with fully trained troops and was now abundantly equipped with modern weapons. This made the location a serious threat to Egyptian communication and supply

Muhammad Adli Kafafi and his colleagues of the REAF's newly created No.8 bomber squadron under *Qa'id Janah* Muhammad Ibrahim Abu Rabi'a, probably photographed immediately prior to Egypt's official entry into the Palestine War. The man on the far right appears to be from the Egyptian Navy, perhaps attached as a liaison officer. (Kafafi family archive)

lines. Once the decision was made to take Nizzanim, the role of the REAF's Tactical Air Force at al-Arish was to be threefold: 1 – Making concentrated preliminary strikes in conjunction with the artillery; 2 – Protecting Egyptian ground forces as they attacked; 3 – Harassing enemy remnants if they tried to retreat.

Mahbub continued:

> The air force successfully carried out the attack on the colony for a full hour, satisfying the first requirement. In reference to the second requirement, waves of aircraft circled around the scene of the operation and watched the routes along which enemy support might come. To implement the third requirement, throughout the operation the aircraft were fully equipped to strafe all those who thought to escape … However, as soon as the infantry advanced to break into the colony it was confronted by mad resistance from the pillboxes, just as the newspapers reported … The second stage was (therefore) promptly started before the first stage had finished. The Egyptians gained significant time and the enemy found himself busy in more than one confrontation … However, once the enemy realised how critical his position had become, the ferocity of his resistance increased.

Lieutenant Mahbub then described the Egyptian commander's decision to abandon the planned slow-but-steady approach in favour of throwing everything into one unrestrained assault. This Mahbub compared with Napoleon Bonaparte's similar decisions at the battles of Lodi and Austerlitz, claiming that the Israeli military leadership fell into confusion because they had only studied modern warfare on the 'sand table'. 'The enemy showed this (confusion) by fleeing into a high building on one side of the colony, which they called the palace, while some of his other elements hid in gardens (perhaps including Abd al-Majid al-Rafa'is' sheaves of wheat) adjacent to the settlement. From there, they rained attackers with a barrage of desperate, reckless fire in an attempt to delay their advance'.

Lieutenant Mahbub now turned to the experiences of a pilot involved, possibly *Tayyar Awal* (First Pilot or Flying Officer) Abd al-Majid al-Rafa'i:

> Although this pilot had not been aware of the developing situation on the ground because it is out of his jurisdiction, he was at least aware of temporarily crisis (sic.) that faced the ground forces. So, he used his experience to find a new solution, which enabled the attacking forces in the purpose of defeating the enemy and the takeover the colony … With some nervousness and hesitation, he dived (sic.) his aircraft over the enemy and strafed them with gunfire. He noticed that the enemy had to duck down to avoid fire opened from the air. This was all that the attackers needed … as the enemy would be forced to duck for periods during which the attackers could infiltrate. Of course, the longer the enemy were keeping their heads down, the better and safer the chances of the attackers. So, his plane, followed by other planes once their pilots realised what their colleague was trying to do, flew high and then one after another swooped on the enemy, opening up with their machine guns … and I am describing what I myself witnessed.

While the battle of Nizzanim was being fought, the Israeli Air Force suffered another loss to enemy groundfire farther north. The aircraft concerned was an Auster (VQ-PAS), which took off from Lydda with Amnon Berman at the controls, accompanied by a female passenger on what is said to have been a photo-reconnaissance mission. It was reportedly brought down by ground-fire over the northern Negev. Although REAF aircraft were operating in the area around the same time, no Egyptian pilot claimed to have downed the Israeli Auster. Even so, there is still some confusion about its identity and the cause of its loss, as the aviation historian Brian Cull informed the author in a letter in 1992: 'The Israeli Auster was shot down, but not by the Fury (prototype) … and definitely carried the registration VQ-PAS, but whether it was the Auster Autocrat, the legal owner of this registration, or another of the two ex-RAF Auster AOP.5s given the same letters, is not clear'. Apparently, the Israelis were using these civilian registrations to sow confusion amongst hostile intelligence agents.

Despite being deeply committed in the struggle for Nizzanim, the REAF carried out one or more raid upon the industrial area of Rehovot, south of Tel Aviv, on 7 June. It was during the latter operation that the Israelis claimed to have shot down an Egyptian Spitfire, though the Voice of Israel only announced this the following day at 1730 hours GMT, and attributed the information the British: 'The BBC announced that during the air battle over Tel Aviv yesterday (7th) an Egyptian plane of Spitfire type was shot down'. The loss was denied by the Egyptians in their Communiqué No.27, broadcast at 2020 hours GMT on the 7th: 'Our Air Force carried out reconnaissance flights over enemy concentrations and supply lines, acquiring important information. Our heavy bombers, escorted by fighters, carried out a successful raid on the settlement of Rehovot, dropping high explosives and incendiary bombs … Our aircraft returned safely from all these operations".

In fact two Israeli Avia S-199 pilots, Maurice Mann and Dov Ben Zvi, who took off to look for the Egyptian intruders, had been given no useful information and were thus unable to find any enemy aircraft. Instead, their presence over Tel Aviv caused the air-raid sirens to be sounded because they were thought to be Egyptian.

Having heard rumours that Egyptian and Israeli prisoners-of-war might be exchanged, Abd al-Rahman Inan remained in a state of uncertainty until sunset on 7 June. Inan then heard many footsteps in the corridor outside his cell, followed by a conversation in English in which one man was talking with an Arab accent. Peering through a small hole in his cell-door, Inan saw Barakat and an Israeli guard in a cell that was being rearranged while the electric light was on. Inan noticed that Barakat was twirling the moustache that now covered half his face. He was also arranging some American magazines on a small table next to his bed, a luxury that Inan had not been allowed. After the Israeli soldier left, leaving the cell door open, Barakat noticed that someone was staring at him through a hole in the opposite cell door. So, Inan loudly sang Barakat's name as if in a song, upon which Barakat showed his fellow prisoner his bandaged knee, indicated that it was a bullet wound.

That night, Inan heard sirens followed by loud knocking on Barakat's door, so he again peeped through the hole in his door.

> Barakat asked the guard to let him go to the (protective) trench, but the guard informed Barakat that this was against the rules. Barakat said that he used to go to a trench in the other (POW) camp when there was an air-raid. So the guard went to consult his leader, then he returned to inform Barakat that every prison has its regulations and that he must stay in his cell.

Eventually one of the guards noticed that the prisoners were communicating through the holes in their doors, so covered them with wood. Thereafter the Egyptians exchanged greeting by coughing, with their guard eventually joining in. Perhaps they had developed a sort of code, similar to that previously used by Barakat and Gamal Irfan Sayf al-Nasr when knocking on the wall between their adjacent cells.

One of the most mysterious events, which reportedly took place on 7 June, was when the captain of a merchant ship called the *Sonja* landed his cargo of spare parts for the REAF 'in error', either in friendly Beirut or hostile Haifa. The vessel may have been the *MV Sonja*, built at Helsingborg in 1940 and operated by the Swedish shipping company A/B Transmarin, but perhaps in Liberian ownership at the time. However, northing more appears to be know and it is possible that this event never even actually took place.

The RIrAF also lost two aircraft on 7 June, one of which was a Hawker Fury fighter of No.1 Squadron. Apparently, the pilot, Captain Hamid al-Dabuni (or Dabbini), was practicing aerobatics over the Mu'askar al-Rashid airbase outside Baghdad when he lost control, probably due to excessive G forces, crashed and was killed. The other was an archaic Gloster Gladiator of No.4 Squadron, based at Mafraq. It seems that Captain Abd al-Sattar Abdullah got lost while flying in bad weather conditions, either on a training sortie or while flying to or from Iraq. He landed the Gladiator safely in the Syrian desert and the plane was discovered after a 15 days search, but Abd al-Sattar Abdallah himself was never found.

At 2050 hours GMT on 7 June, Syrian Communiqué No.23 reported that: 'A squadron of our aircraft attacked the settlements of Nahariya, Hanita and Massawiyah (unidentified). Enemy fortifications were destroyed and heavy losses inflicted … Another squadron attacked the settlements of al-Nabi Yusha' and Harrawi with bombs and machine-guns, causing panic and terror among the enemy'. Clearly the SyAAF Harvard units were still hard at work, despite their limited numbers of aircraft and aircrew. Meanwhile, strenuous efforts were continuing outside Damascus to complete the advanced training of young pilots such as Faisal Nassif. His diary was nevertheless still mainly concerned with the difficulties he and his young wife, Alice, were facing: '7th June: I took her back to al-Mizzah (the main SAF airbase near Damascus) and now we live with Lieutenant Mulhab'.

Encouraged by their success on the previous day, Lebanese troops pushed forward to capture Ramot Naftali and Qadas on 7 June, potentially opening a route to the Hula Valley. Sadly, 7 June witnessed yet another alleged Israeli war-crime against Palestinian civilians.

8 JUNE 1948

GROUND OPERATIONS

Tuesday 8 June was a relatively quiet day on the ground, though far from uneventful in the air. On the southern front Egyptian and Israeli forces consolidated their positions and established a moderately stable if ill-defined front line from the coast north of Isdud to Bayt Jibrin in the Judean foothills. This formed the northern edge of the so-called Egyptian isolation belt, separating the main Israeli forces in central Israel from those in the Negev. Farther north, the inhabitants of most of the Palestinian Arab villages had already been expelled by Zionist or later Israeli forces.

On the Syrian and Lebanese fronts, the Israeli Army rushed a battalion of the Carmeli Brigade to the area following the Lebanese successes on the 7th. Even so, it would not be long before the Syrians regained their momentum, again threatening to break through the Hula Valley.

An Egyptian Army radio post in Palestine. With both sides using, essentially, the same equipment – usually of British- or US-origin – radio stations were widely deployed not only for communication with friendly forces, but for listening to enemy communications, too. (David Nicolle collection)

AIR OPERATIONS

A reflection of current British policy can be seen in secret British Air Ministry cypher messages on 8 and 9 June 1948. These confirmed Air Ministry signal MSX.305 of 1 June 1948, stating that supplies set aside for delivery to the REAF currently at No.107 Maintenence Unit in Kasfarit were to remain suspended indefinitely. The REAF would have to do the best it could with the material already in its hands. In fact, during the first week of June, the REAF reportedly sent several of its outdated Spitfire 5s to al-Arish, perhaps indicating how difficult it was to keep the slightly younger Spitfire 9s serviceable.

For a while, the Voice of Israel began adopting an almost hysterical tone in its news broadcasts. It reflected the pressure that the Israelis were under, not only within the military, but also more particularly amongst civilians. This was heard in a continuing tendency to blame the British government for things that were no longer under British control. The news broadcast at 1730 hours on 8 June is one of the more strident examples, reporting that:

> Nuqrashi Pasha (the Egyptian Prime Minister) has sent his corrupted British planes to bomb Tel Aviv indiscriminately. In this barbaric Egyptian air attack, these planes dropped their bombs from a very high altitude, resulting in injury to peaceful passers-by. Nuqrashi Pasha has given orders to his pilots to avoid coming into contact with Jewish pilots, because he knowns that Egyptian pilots will certainly be killed if they dare to fight our pilots.

This broadcast came at the end of the day that had seen the first confirmed, and highly visible, clash between Israeli and Egyptian fighter aircraft above Tel Aviv. According to an Egyptian communiqué broadcast at the usual time of 2200 hours GMT:

> The Air Force made several concentrated attacks on the port of Tel Aviv. The dive bombers bombed the quays and harbour warehouses of Tel Aviv, setting them on fire. During this attack our aircraft met three Zionist fighters, which our planes attacked, causing them to take evasive action after hitting one of them, which was probably unable to reach its base. Our aircraft … observed a cargo vessel, some steam boats and some barges, which they attacked with high explosives. The cargo ship received a direct hit after which it set course for the open sea. Some big patches of oil were observed. Some of the barges were sunk. During their repeated (later) attacks our aircraft observed that fire was still taking hold in the port district as a result of the first raids. Our aircraft returned to their bases safely.

The REAF Tactical Air Force Operational Diary, which provides details about sorties from al-Arish and later also from new satellite airfields in north-eastern Sinai, finally becomes available on 8 June 1948. A small number of pages remain missing and this valuable source of information does not include REAF sorties from aerodromes west of the Suez Canal. Nevertheless, it clearly shows that the wording of official Egyptian communiqués concerning REAF activity, broadcast earlier in the conflict, were based directly upon this and presumably also other Operational Diaries. Only later did any significant degree of journalistic elaboration creep into official communiqués broadcast from Cairo.

The first mission on the first page of the available section of the REAF's al-Arish Operational Diary is worth quoting in full, showing that it was organised in the same manner as British RAF operational records upon which the Egyptian version was based.

> Sortie Number: 1.
> Target: Warehouses of Tel Aviv, stores and jetties of Tel Aviv.
> Weapons: HE bombs and incendiaries.
> Serial numbers (of aircraft involved): 6-- (illegible) & 654.
> Aircrew: Mustafa al-Utaify, al-Tahir (Tahir Zaki).
> Aircraft type: Spitfire – (illegible but presumably Mk. 5) & Spitfire 9.
> Date: 08-06-48.
> Takeoff: 05.25 & 05.25 (on this occasion both machines took to the air at the same time, though this would not always be the case).
> Landed: 06.25 & 06.25 (or 06.45 which is unlikely).
> Visibility: Good.
> Enemy defences: 2 Messerschmitt and 1 Norseman.
> Remarks: Aeroplanes carried out a successful raid over warehouses and Tel Aviv Port berths. Two hostile planes intercepted our planes but our planes executed their mission successfully and hostile planes escaped (translation not entirely clear here). Our aircraft attacked stores and jetties of Tel Aviv and before (they) reach (this target?), 2 Messerschmitt and 1 Norseman attack our aircraft but it (the Egyptian aeroplane or aeroplanes) hit & attack Israeli aircraft and they ran away.

This sortie by a pair of Spitfires was the first of three during the morning, but was the only one to confront enemy aircraft. When interviewed many years later by the author, Tahir Zaki remembered the combat well, but also seems to have conflated this mission with another, particularly where targets were concerned. In Tahir Zaki's own words:

> Another successful attack which I made with al-Utaify was against Tel Aviv. We were at 6,000 feet when Utaify suddenly turned to the right in a sort of evasive action. He hadn't actually seen the two Israeli aircraft which were about to open fire on us; it was more like an instinctive action. The Israelis only made one pass before flying away. My aircraft was hit in the wing but it was still flying OK. I looked down and saw a big hotel on the shore. It was surrounded by Army cars so I attacked this hotel and made accurate hits. Later we found that there was a military conference taking place and my two 250lb bombs fell right inside the hotel. During the afternoon Israeli radio said that many senior officers had been killed, some of them commanders.

It was subsequently claimed that Egyptian intelligence learned that senior Israeli military commanders were in conference in the Allenby Hotel. According to al-Tahir's interview with an Egyptian journalist, published in *Al-Tabar* on 2 June 1998, under the title 'The pilot who attended his own funeral', he named the targeted hotel as *Al-Nabi* on Allenby Street. In fact, the *Yarden Beach Hotel* had been requisitioned by the Israeli Air Force for use as an HQ during the Palestine War. It was not on the beach, though close, and was near Allenby Street. The only air raid that is known to have killed senior Israeli officers had taken place on 29 May – unnamed Egyptian pilots striking an Israeli HQ in Ramat Gan on what was then the eastern edge of Tel Aviv.

Mustafa Salah al-Din al-Utaify would eventually rise to the rank of Major General in what became the United Arab Republic Air Force, serving with distinction in all Egypt's air conflicts up to and including the October 1973 War. He was a mentor, hero and friend of Mustafa Shalabi al-Hinnani who subsequently wrote that the Israelis offered a reward of 30,000 Egyptian pounds (then the equivalent of £29,250 sterling, and in 2021 worth over a million) to anyone who could kill al-Utaify. True or false, this story certainly shows the high regard in which Mustafa Salah al-Din al-Utaify was held by his Egyptian colleagues.

The Israeli pilots' accounts of this combat were, of course, different. Early in the morning of 8 June, two Avia S-199s were ready to fly a patrol south of Tel Aviv, Egyptian Spitfires having been in action in this area the previous day. The patrol was to be led by Modi Alon, with Giddy Lichtman as his wingman. After some 15 minutes in the air, Modi Alon suddenly rolled into a dive, having seen four Spitfires heading towards Tel Aviv from the south. This was possible, though very unlikely, as the second REAF sortie consisting of a pair of Spitfires took off at 0610 hours, while Sortie No.1 returned to al-Arish at 0625 hours.

When Alon opened fire, the supposed four Egyptian Spitfires split in four different directions. Lichtman followed a Spitfire that headed south. In Lichtman's own words: 'The Egyptian pilot evidently saw me, for he started to do some real half-arse manoeuvres. A roll, a steep turn and then he rolled the aircraft onto its back, pulled the stick back and dived towards the ground'. After further manoeuvring, Lichtman got behind the Egyptian and opened fire, claiming to have seen pieces fly off the Spitfire's wing. Convinced that the Spitfire was doomed, Lichtman flew home. He was subsequently credited with downing the first Arab fighter to fall to the Israeli Air Force, an

Israeli ground observer having reported that the aircraft went down 'in flames' south of Tel Aviv.

No other accounts refer to the Norseman aircraft mentioned in the REAF Operational Diary, while an official report published in the Egyptian press may have confused this day's events with what happened a few days later: 'During this attack our aircraft met three Zionist fighters, which our planes attacked, causing them to take evasive action after hitting one of them, which was probably unable to reach its base ... Our aircraft returned to their bases safely'.

As mentioned above, an early account of Modi Alon's life suggested that three days after he was credited with downing Egyptian Dakotas, he attacked some REAF Spitfires. He reportedly downed one but was shot up by another, supposedly landing at his home aerodrome slightly injured. The events of 8 June fit this account much better than those of the 6th, but this account would subsequently be denied.

According to the REAF Operational Diary, the two Spitfires of Sortie No.2 took off from al-Arish at 0610 hours. Muhammad Abu Zaid was at the control of a Spitfire 9 number 633, Wahdan flying a Spitfire 5 number 615. Carrying the same weapons as Sortie No.1, they aimed for the same targets. Visibility was good and no enemy defences were encountered and according to the comments column: 'Our planes executed a successful raid over the same targets and bombed these targets which broke out into large fires with smoke observed rising to 500 feet (approximately 152m). A boat of 200 tons was observed outside the port along with 3 small boats heading from it towards the shore'. Wahdan landed back at al-Arish at 0720 hours, perhaps covered by Abu Zaid because Israeli fighters were known to be in the air, before Abu Zaid landed five minutes later.

Attacks on the same targets in and around Tel Aviv harbour were continued by Sortie No.3. It was again led by Squadron Leader Abu Zaid, in Spitfire 9 number 634, accompanied by Muhammad Safwat in Spitfire 9 number 646, taking off at 0940 hours and landing at 1100 having faced no opposition. The summary of this final sortie of the morning merely stated: 'Successful bombing of the port berths and warehouses, with a big fire and black smoke and --- (illegible but probably some form of ship) hit beside the jetties'.

An Israeli account of these morning raids reported that: 'The remaining Egyptian Spitfires (one having believed to be shot down) carried out dive-bombing attacks on Tel Aviv which caused the deaths of nine civilians and injuries to 45 others'. At 1645 hours GMT, the Voice of Israel somewhat contradicted this by stating that: 'Egyptian aircraft dropped bombs indiscriminately from a great height. Five persons were killed'.

By then the REAF's afternoon sorties were already under way farther south, Sortie No.4 being a reconnaissance mission undertaken by REAF Lysander number 516, flown by Ra'uf with no mention of an observer-gunner. His task was to photograph Israeli settlements or positions north-east of what remained of the Palestinian villages of Qastina, Yasur and an unidentified location written as Bayt Dawud. Ra'uf took off at 1630 hours, returning at 1800 hours having taken pictures north-east of a line which included another location written as Bawin. Sortie No.5 consisted of a pair of REAF Spitfire 9s, flown by Galal in number 634 and Yahya al-Aidarus in number 642. They took off ten minutes after the Lysander, gave cover to the former and attacked Israeli positions in the same area before returning ten minutes apart at 1815 and 1825 hours. The visibility was still good and the Egyptians pilots faced no opposition.

The purpose of these two missions nevertheless remains unclear because Israeli settlements in this area were not officially established until 1949. However, the Arab inhabitants of Yasur village were expelled the following day, following an assault by the Israeli Givati Brigade, while the comments column for Sortie No.5 specified that the Spitfires attacked Yasur 'colony': 'High explosive bombs fell inside the colony, striking the building, and a building was seen collapsing. Also, planes investigated and protected (Egyptian) front unit vehicles'.

The reference to Egyptian troops perhaps indicates that the fate of Yasur still remained in the balance or that the village had only just fallen into Israeli hands, while the need to send a highly vulnerable Lysander on a photo-reconnaissance mission indicates that the Egyptian Army was desperate for accurate information from this locality.

The only other Arab air operations on Tuesday 8 June were carried out by the SyAAF. Syrian Communiqué No.24 concluded the day at 2050 hours GMT by reporting that: 'Our aircraft bombed successfully the settlements of Harrawi, Hanita and al-Nabi Yusha', inflicting heavy damage, and destroyed enemy fortifications which sustained heavy losses. A group of our aircraft attacked the settlement of Najmat al-Subh (Ayelet HaShahar) with high explosive and machine-gun fire, registering direct hits; they destroyed enemy establishments and set other places on fire'. Meanwhile, Syrian pilot Faisal Nassif's diary still focused on domestic matters: '8th June: We have rented a room in the same house where Lt. Mulhab Iyd lives (in the Damascus suburb of al-Mizzah, next to the aerodrome). We bought a bed and my mother has sent us a blanket, a pillow and so on'.

The following night, 8-9 June 1948, while the alternative road to Jerusalem was still under construction, Haganah command made a further attempt to take Latrun before the UN Truce came into force. The Yiftach Brigade was therefore sent from Galilee to the Jerusalem Corridor. This plan called for a coordinated attack by three Palmach battalions: one from the Harel Brigade and one from the Yiftach Brigade were to attack Latrun from the east, with the Harel battalion taking Yalu Hill to the east of Latrun. Israeli control of this elevated location would enable the Yiftach battalion to advance through Harel positions to attack Guns Hill. Meanwhile, another Yiftach battalion was to attack Biddu to cut off Arab Legionionaries defending Radar Hill.

After a number of delays, the operation began at midnight but the Harel battalion missed its target and instead attacked Guns Hill, where it faced a stubborn resistance. A few elements of its force occupied part of the hill, but before dawn the Arab Legion launched a determined counter attack, recovering its lost positions. When the Yiftach battalion advanced towards what it believed a friendly position, it came under intense hostile fire and this third Israeli attack broke up in disarray.

9 JUNE 1948
GROUND OPERATIONS

As the proclamation of a UN Truce seemed imminent, the Israelis were eager to improve their position facing the Egyptians and to recapture Nizzanim. First, a Givati force occupied Hill 69 (so-named because of its indicated elevation above sea level), 2km north-east of the settlement. This was to be used as a base for a subsequent offensive. Simultaneously, the 7th Negev Battalion mounted a diversion attack against Isdud, but this was no more successful than its predecessor. Elsewhere, other elements of the Givati Brigade took full control of what remained of the Palestinian villages of Yasur and al-Masmiyah al-Saghirah (also called Masmiyah al-Huraniyah), though the defenders of neighbouring al-Masmiyah al-Kabira held out until July.

AIR OPERATIONS

Most REAF sorties on the 9th were in response to the Israeli counter-offensive. Even an Egyptian attack upon Tel Aviv may have been conducted in the hope of diverting Israeli attentions away from the front. While the intensity of Arab air activity was sustained, and to some extent increased, the tone of Egyptian official communiqués broadcast from Cairo started to exaggerate the scope, intensity and effectiveness of REAF air-strikes. Thus, a broadcast at 2000 hours GMT on 9 June claimed that:

> Our aircraft strafed an enemy convoy south of Tel Aviv, inflicting considerable losses in men and destroying a number of trucks. Our aircraft raided the settlement (or now enemy-held village) of Yasur … Our heavy bombers carried out repeated and sustained raids on the Rehovot district. High explosive and incendiary bombs were dropped on the chemical factory and the surrounding buildings, setting them on fire … Two Zionist Messerschmitt planes tried to engage our aircraft but one of them was shot down in flames. Our aircraft spotted an enemy Messerschmitt plane, during their return journey from one of the raids over the Aqir airfield. The aircraft was attacked and brought down on the airfield. Our planes returned safely to their bases.

Though Cairo's broadcasts started to exaggerate, the tone of the REAF's al-Arish Operational Diary remained restrained and down-to-earth. Thus Sortie No.1, which was intended to be a solo, armed reconnaissance by Flight Lieutenant Mukhtar Mahmud Sa'id, in old Spitfire 5 number 607, took off at 0515 hours but had to return 35 minutes later because of very poor visibility.

Sortie No.2 targeted the warehouses and docks or jetties of Tel Aviv port. Spitfire 9s numbers 651 and 633, flown by Muhammad Abu Zaid and al-Qarmuti, with Spitfire 5 number 616, flown by Musilhi, each carrying high explosive and incendiary bombs, took off between 0810 and 0815 hours. Visibility was now good and the enemy offered no reported opposition, the three Spitfires returning to al-Arish between 0925 and 0930 hours. The comments on this sortie suggest that it was visibly successful:

> Bombs fell on a petroleum depot. It was directly hit and caught fire, large flames being seen to a height of 700ft (over 200m). Some bombs fell near the southern port where stores caught fire and during the aircrafts' return (they) attacked vehicles heading from Tel Aviv to Rishon-Le-Zion. Our aircraft struck four vehicles and (also) found the boat hit yesterday aground near the shore off the pier.

Sortie No.3 was a repeat of Flight Lieutenant Mukhtar Sa'id's earlier aborted armed reconnaissance in a Spitfire 5 number 605. He took off at 0925 hours and returned at 1100 hours, having completed his observation of enemy positions north of Isdud. Sa'id, presumably on his way back, also took photographs of the recently lost village of Yasur, the Israeli settlements of Gat, Dorot, and other unidentified locations.

The REAF now launched three offensive sorties from al-Arish. Sortie No.4 targeted Gat (Kibbutz Gat rather than larger town of Kiryat Gat, which would be founded in 1954). Two Spitfire 5s, numbers 601 and 615, were flown by Yahya al-Aidarus and al-Tarazi, taking off quarter of an hour after Mukhtar Sa'id landed. They dropped HE and incendiary bombs inside Gat before landing safely at 1300 hours, having noticed no enemy opposition. The next two sorties hit the Israeli industrial town of Rehovot. No.5 consisted, as usual, of a pair of Spitfires, this time Mk.9 machines – as they were more likely to face opposition from enemy aircraft. Number 633, flown by al-Qarmuti, and number 646, flown by Galal, took off to together at 1555 hours. Visibility was good and in the words of the comments column in the Operational Diary: 'Two aircraft hit the Chemical Factory where green and white smoke was observed. Bombs fell directly on two factory buildings'.

Unfortunately the next sentence in the available photocopied poor-quality carbon-copy of the Operational Diary is barely legible, but it is clear that the Egyptian aircraft were intercepted by a reported two Israeli 'Messerschmitts' (Avia S-199s) and, after exchanging fire, both Spitfires escaped. One interpretation of this sentence might claim that an enemy machine was shot down, while another merely states that 'we survived'. Galal landed back at al-Arish about half an hour after al-Qarmuti, and as both Spitfires would fly sorties on the next two days it is clear that neither was damaged.

Israeli sources remain remarkably quiet about these six Egyptian sorties. However, one Avia S-119 was reportedly damaged by ground-fire while being flown by Modi Alon on the following day. No claim to this effect was made by a relevant Arab source, so might this damage have been suffered in combat with REAF Spitfires on the 9th?

Sortie No.6 intended to repeat No.5, by again attacking Rehovot. Another pair of REAF Spitfires, of unknown Mark because their serial numbers are not mentioned, took off at 1630 hours, before al-Qarmuti or Galal had returned. At their control were the more experienced Squadron Leader Muhammad Abu Zaid and Flight Lieutenant Mukhtar Sa'id. After bombing Rehovot with HE and incendiaries, Abu Zaid and Sa'id were probably informed by radio of the attack on their colleagues by Israeli fighters, because they now turned towards the major Israeli aerodrome of Aqir. Here they described the enemy's defences as 'Stronger over Aqir airfield'. The comments column of the Operational Diary merely states: 'We hit colony by bombs and we saw smoke about 1000 feet (over 300m) and at that time one of our aircraft saw --- (unclear) at Aqir airbase'. Abu Zaid, Mukhtar Sa'id and Galal landed back at al-Arish with only 5 or 10 minutes between them, having flown back together.

Both Abu Zaid and Mukhtar Sa'id were to be lost later in the Palestine War and both would be celebrated for their clearly different characters. Flight Lieutenant Mukhtar Mahmud Sa'id was remembered for his cheerfulness, for example, and for describing his aircraft as 'King Faruq's pride and the mount of his knights'. Once again Egyptian radio made events over Rehovot more epic: 'Two Zionist Messerschmidt planes tried to engage our aircraft but one of them was shot down in flames. Our aircraft spotted an enemy Messerschmidt plane during their return journey from one of the raids over the Aqir airfield. The aircraft was attacked and brought down on the airfield. Our planes returned safely to their bases".

Effectively ignoring air activity over the southern front, the Voice of Israel focused instead upon the IDF/AF's activities in the central region where, according to a broadcast at 1730 hours GMT: 'Israeli aircraft bombed fortifications and a radar (or radio?) station at Biddu and al-Nabi Samwil scoring direct hits. Enemy guns there, which were shelling peaceful districts of Jerusalem, were silenced. The Israeli Air Force also bombed enemy lines at Ras al-Ain (probably the future site of Rosh HaAyin adjacent to Majdal Yaba) and Majdal Sadiq (Majdal Yaba)'.

At 1830 hours GMT, the Voice of Israel reported that: 'Enemy planes attacked Zikhron Ya'akov (south of Haifa), dropping bombs in the streets. Women and children were killed by machine-gun fire'.

Here the enemy aircraft might have been Iraqi. However, the Iraqis were even more silent about the RIrAF on 9 June. In contrast, at 2050 hours GMT, Damascus broadcast Syrian Communiqué No.25, reporting somewhat optimistically that:

> Our aircraft bombed mobile enemy columns which were trying to attack the Arab village of Lubiya (west of Tiberius). Several armoured cars were destroyed by bombs and a number of armoured cars were damaged by fire from machine-guns. A squadron of our Air Force bombed enemy concentrations in the neighbourhood of the two Arab villages of Lubiya and (the neighbouring hamlet of) Miskanah, on the road between Tiberius and Nazareth, thus frustrating the enemy's plans. Enemy concentrations were dispersed and losses inflicted.

The Israeli version of these two SyAAF sorties confirmed that they took place in the afternoon. In the first, one Harvard supposedly dropped two bombs on an unnamed Israeli village in Galilee, without causing reported damage, while the second took place in the evening when four Syrian Harvards flew low over an unnamed kibbutz and dropped three bombs which caused no damage. Apparently, someone on the ground claimed that a Harvard suffered mechanical difficulties and was unable to drop its load; instead returning to its base with the bombs still attached. Back in Damascus, Lieutenant Faisal Nassif had yet to be sent to a front-line squadron. Instead, his diary recorded that he and his wife borrowed a table, four chairs and some other illegible object from the Syrian Officers' Club.

10 JUNE 1948

GROUND OPERATIONS

Thursday 10 June saw the opposing armies making some significant moves. During the previous night, the Israeli 2nd Givati battalion, which had been supposed to retake Nizzanim from the Egyptian, lost its way and so the assault was called off. But, it could still hope to win control of Hill 69, along with the villages of Jusayr and Julis plus an important nearby road junction, which were all still held by ill-equipped and barely trained Arab irregulars. The villages were taken, but that was all. Much farther south the Israeli seizure of an ex-British police fort at Bir 'Asluj, overlooking the road between Beersheba and the Egyptian frontier, also achieved virtually nothing because Egyptian Army engineers quickly laid a new road bypassing the fort and thus re-establishing Egyptian lines of communications. Furthermore, the Bir 'Asluj fort would soon be regained by the Egyptians.

Extracts from another little-known but interesting Arab military diary are available from this date onwards. It belonged to Lieutenant Salah Badr, of the Egyptian Infantry. Unlike Syrian Harvard pilot Faisal Nassif, Salah Badr would survive the war, his diary being taken by the Israelis when he was a prisoner. Like Nassif's diary, it is a very personal document, and while Nassif was more often than not concerned with domestic affairs, Badr seems to have dedicated much of his diary to how life on the front-line was affecting his digestive system. On 10 June, however, Salah Badr was still fit, recording that he and a colleague named Abd al-Qadir Hassan went by overnight train from Egypt to Palestine, Badr having been ordered to join the 9th Infantry Battalion.

Higher up the chain-of-command, General Naguib was not ready to allow the Israelis to control events in the Negev. On the 10th, Egyptian forces therefore took up positions near the junction where the lateral road, controlled by the Egyptian Army, crossed the main inland road from Tel Aviv to the Negev settlements. By doing so, they dominated the Israelis' only viable supply route south.

However, the Israelis were equally determined not to allow the Egyptians to control the situation. When Israelis troops returned to the lost settlement of Nizzanim before dawn, they came under intensive Egyptian fire. Their leading platoon penetrated the

The police fort of Bir 'Asluj. (David Nicolle collection)

settlement only to be quickly driven out, after which the entire attacking force was forced to retreat. Egyptian artillery fire pursued them as they withdrew to Hill 69, where a substantial force had already been assembled. Shortly afterwards two Egyptian infantry companies, supported by a light tank company, attacked Hill 69 where Israeli resistance promptly collapsed, followed by a disorganized retreat. As a consequence of this abortive attempt to recapture Nizzanim, Israeli casualties reportedly rose to more than 300 killed, in addition to 10 prisoners, while the Egyptians suffered just 5 killed and 13 wounded.

Other Givati Brigade units were sent to expel the inhabitants of several Arab hamlets between Yibna (taken on 4 June) and the Israeli settlement of Be'er Tuvia. The Givati Brigade completed most of its mission with little difficulty because none of these villages had yet been garrison by Egyptian troops, and were still being defended by Palestinian militias. Their inhabitants fled either to Falujah or towards the Egyptian-held coast. Givati also took Hill 113, located 1km north of the strategic road junction, and apparently only defended by a handful of Egyptian soldiers – but not the road junction itself. Givati was also expected to capture Julis, where Israeli forces had control of the ex-British barracks on the night of 27-28 May. This village eventually fell on 11 June, (probably within hours of the UN Truce theoretically coming into effect). Jusayr, lying 4km nor-nor-east of Falujah was also to be cleared, though this battered frontline village would not fully fall into Israeli hands until 17 July, when its remaining Palestinian inhabitants were expelled, along with those of Hatta which fell at the same time.

In order to improve the Israeli position before the UN Truce and the arrival of UN observers, the Negev Brigade was ordered to capture Iraq Suwaidan's police fortress, which dominated the lateral road 1km west of Iraq Suwaidan village. More importantly, it also overlooked the strategic north-south and east-west road junction 2km farther west. This, the Israeli troops failed to do during a night assault because a company of Egyptian infantry now garrisoned the fort. However, Negev Brigade units were able to establish positions in the outposts of the Arab villages of Kaukaba and al-Hulaiqat that respectively lay 2.5 and 6km to the south of the junction on the road to the Negev.

On the northern front, the Syrian 2nd Brigade mounted a successful attack on 10 June, as a result of which the Syrian Army controlled all the river crossings east of Mishmar HaYarden. Working in close cooperation with infantry, Syrian light tanks had pushed forward and overran the outer defences of the settlement. A fierce battle then took place and, at noon, with most of the Israeli defending force dead and the rest taken prisoner, Mishmar HaYarden fell into Syrian hands. As a diversion, two other Syrian units again attacked the Israeli settlement of Ein Gev, on the eastern shore of Lake Tiberias. They succeeded in diverting two Israeli brigades, but because the Israelis hurried in reinforcements, the Syrians failed to capture these two settlements. Sadly, 10 June was again marked by yet another alleged Israeli war crime against the Palestinians.

AIR OPERATIONS

With a UN Truce fast approaching, the REAF was busy on 10 June. However, the al-Arish Operational Diary only listed tactical raids by Spitfires, referred to in Egyptian official Communiqué No.30 broadcast from Cairo at 2230 hours GMT as fighter-bombers. The aircraft, which this communiqué also reported attacking the Tel Aviv area, must therefore have been REAF Dakota bombers operating out of Almaza or Cairo West.

According to this communiqué:

> Our aircraft raided the Tel Aviv area, dropping bombs on a petroleum (storage) plant, which caused fires rising to a height of 700ft (over 200m). At Tel Aviv port the warehouses were bombed and set on fire. Our fighters bombed the villages of Yasur (taken by the Israelis on 9 June), al-Qubab (taken by Israeli forces on this day) and Gat (the Israeli settlement), destroying some stores, and strafed troop concentrations in the settlement causing heavy losses in men. Our aircraft raided the towers of enemy fortifications on the road between Jerusalem and Latrun and the region of Bayt Mahsir (taken by the Israelis on 10 May), causing considerable damage. Our planes returned safely to their bases from all these operations.

The first REAF sortie from al-Arish on 10 June was an armed reconnaissance of enemy concentrations in Yasur, using HE and

A Humber armoured car of the Egyptian Army, as seen underway down a minor road in southern Palestine in June 1948. (David Nicolle collection)

incendiary bombs, carried out by a pair of Spitfire 9s, numbers 642 and 646, flown by Mustafa al-Utaify and Yahya al-Aidarus. They took off at 0730 hours, reported moderate visibility and no hostile fire, landing again at 0825 and 0830 hours. The comments column in the Operational Diary remarked that: 'Yasur settlement was bombed and the bombs fell on warehouses which partially collapsed. Also made a reconnaissance of the area east of the (Egyptian held) Nizzanim settlement'. Sortie No.2 again targeted enemy troop concentrations around Yasur, Spitfire 9 number 651 being flown by al-Qarmuti, and Spitfire 5 number 616 by al-Wasaimi. After taking off between 0825 and 0830 hours, around the same time that Sortie No.1 returned, the Egyptian pilots reported moderate visibility with low clouds, no visible enemy resistance, and struck what they described as small groups of enemy troops.

Sortie No.3 was mounted in support of Arab Legion troops west of Jerusalem, by finding and attacking enemy concentrations at Bayt Mahsir overlooking the vital road linking Jerusalem to the coastal plain. Both aircraft were Spitfire 5s, numbers 601 and 615, flown by Muhammad Abu Zaid and al-Abadi respectively. Taking off at 1245 hours and returning at 1355 hours, they found poor visibility although the enemy did not apparently fire at them. The comments on this sortie simply stated that: 'Towers (water?) on a hill in the settlement were hit. On the Jerusalem-Latrun road a lorry was strafed near Bad al-Wad'. Abu Zaid and al-Abadi had presumably also been ordered to look at the area west of Ruhamah, in the northern Negev, on their way home, as they are reported that: 'Two armoured vehicles were strafed to the north of Dorot and the area was explored, and the findings were submitted to the Army'.

The two Spitfires of Sortie No.4 took off at 1245 hours, flown by al-Qarmuti and al-Wasaimi. The types are unknown, as their serial numbers were not given in the Operational Diary. Both landed at 1355 hours, reporting good visibility and no hostile fire, having bombed the Israeli settlement of Gat. However, no enemy were visible and their bombs caused no fires.

At 1335 hours, Sortie No.5 targeted unspecified settlements in the northern Negev. Other evidence suggests that these may have been in the area of Nir Am and Dorot. The aircraft were Spitfire 9 number 633 and Spitfire 5 number 606, this seemingly now an established pattern, flown by Mustafa al-Utaify and Wahdad. Visibility was again good, no hostile fire was reported, and according to the pilots' debriefing: 'The east of the settlement was bombed and no movements were observed near (east of) the coastal road between al-Majdal and Gaza'.

To the north, SyAAF Harvards were similarly in action on Thursday 10 June, continuing to attack Mishmar HaYarden in support of a Syrian Army assault during which around thirty 'armoured vehicles' crossed the river Jordan and surround Mishmar HaYarden. Syrian infantry then fought house to house to take the settlement.

According to Syrian Communiqué No.26, this was a very successful day:

> An enemy aircraft flew over our lines but was met with concentrated anti-aircraft fire which forced it down in the area of Rosh Pina … In all these battles our infantry and mobile and air forces cooperated … The Air Force carried out continuous raids on the enemy from dawn till evening, successfully attacking military objectives in the settlements of Rosh Pina, Najmat al-Subh (Ayelet HaShahar), Dafna, Dan, Ein Gev and – (unclear), as well as concentrations of enemy troops and armoured cars which the enemy was massing for a counter-attack on positions which we had previously occupied. All our aircraft returned safely.

Israeli sources stated that Syrian Army operations were supported by six Harvards on 10 June; these repeatedly flew over Rosh Pina and Ayelet HaShahar. Nevertheless, the SyAAF's shortage of serviceable machines meant that other Syrian assaults on Dan, Ramot Naftali, Dafna and Kfar Szold lacked air cover, despite urgent requests from infantry commanders who were overheard by their Israeli counterparts. Later in the day, the six available SyAAF Harvards supported another Syrian attempt to take Ein Gev, but this time without success as the Syrian infantry were only accompanied by a single armoured car. A further assault the following night, this time without air support, also fell back. A similar Syrian night attempt against Ramot Naftali failed, according to a Voice of Israel broadcast at 1030 hours GMT the following day.

It was fortunate for the SyAAF Harvard aircrew that the Israeli Air Force was currently only using its Avia fighters against the Egyptians. Meanwhile additional Syrian aircrew, probably including those who had recently completed their initial training in Iraq, continued to train at al-Mizzah near Damascus. Here their instructors included the Croatian pilot known to his pupils as Mato (Mato Dukovac). For Faisal Nassif, who was nearing the end of his training, the final days of the First Phase of the Palestine War seemed far away, as he recorded in his diary entry for 10 June: 'Our life is quiet, we (he and his wife Alice) are both happy. She loves me and I love her and I pray to God that our happiness will last'.

On the night of 10-11 June, with the First UN Truce approaching, the Israelis decided to try yet again to take Iraq Suwaidan and its police fort. The Negev Brigade was given this task but would fail because the place was now held by Egyptian regular troops. When Israeli troops launched a ferocious attack, sappers from the Negev Brigade breached the three barbed wire fences which had they expected, dismantling mines as they went. The sappers then also had to breach an unexpected fourth fence. Throughout, they had been under sustained fire from the fort's defenders who sent up flares to illuminate the attackers. A wave of Israeli infantry was then thrown against the fort, but came up against a fifth barbed wire entanglement and had to retreat under heavy fire.

The story of this Egyptian victory would be widely discussed within the Egyptian Expeditionary Force, and junior officer Salah Badr later wrote about it in his diary entry of 7 July. He proudly maintained that the fort had been built by an Arab contractor (though working for the British, to a British design), that this 'block of concrete' menaced the Israeli settlements to the north and south. He believed seven enemy assaults lay broken at its feet, even those supported by armoured half-track vehicles, while a wooden memorial plaque had been erected in memory of two Egyptian soldiers killed in its defence. To the Israelis, this formidable fortification simply came to be known as the 'The Monster on the Hill'.

Also on the night of 10-11 June, the Israeli Air Force dropped bombs on the outskirts of the Syrian capital, Damascus. According to Syrian Communiqué No.27 broadcast at 1845 hours GMT on the 11th: 'A hostile aircraft flew over Damascus early this morning (actually late in the night of 10th-11th) and dropped a number of explosives and incendiaries on the suburbs of the town, but was driven off by anti-aircraft guns. No losses in material or men were reported'. The Israeli version, announced on the Voice of Israel at 0400 hours GMT, shortly after the raid took place, stated that: 'Aircraft of the Israel Air Force attacked military objectives in various parts of Damascus during the night. Our aircraft flew over

Syrian infantry as seen during training, sometime in 1948. Notable is the absence of helmets. Before the outbreak of the Palestine War, many Syrian troops wore a version of the French Adrian M26 helmet, dating from the beginning of the Second World War. (David Nicolle collection)

the city for 15 minutes. All our aircraft returned safely'. A subsequent Voice of Israel broadcast at 1730 hours GMT was more triumphalist in tone, maintaining that: 'In an Arabic transmission, Haganah (the earlier Zionist paramilitary army which had become the Israeli Defence Force on the creation of the State of Israel) stated that the enemy pilots did not dare to attack the Israeli planes over Damascus'. In reality the SyAAF had neither suitable planes nor trained pilots for a night interception.

This raid had actually been carried out by an Israeli Dakota (perhaps an ex-Air France aeroplane) which had been more or less converted into a bomber, though, unlike the Egyptian Dakota "bombers" which had bomb-racks beneath their wings and fuselage, in this Israeli version a team of six bombardiers reportedly rolled two tons of bombs out of the cargo doors. It is therefore surprising that Faisal Nassif's Diary makes no mention of the Israeli air-raid, despite living so close to al-Mizzah aerodrome. His entry for 11 June merely recorded that: 'My brother and my mother often visit us. They are very kind to me and I did not expect from them such an interest in Alice'.

According to the Voice of Israel news broadcast at 1030 hours GMT on the 11th: 'Last night our planes bombed enemy concentrations at Ramlah'. Sadly, the Israelis were accused of yet another atrocity against Palestinians on the night of 10-11 June.

11 JUNE 1948
GROUND OPERATIONS

The Truce announced by the United Nations would last for 28 days. The Egyptian government had initially opposed its acceptance, but King Abdullah of Transjordan, whose influence amongst the Arab leaders remained considerable, argued in favour. Transjordan's Arab Legion had, of course, already achieved most of its war aims. The Egyptian Army's two-pronged advance, which was largely confined to Arab areas of Palestine as designated by the UN Partition Plan, now controlled most of the Gaza region, Beersheba and Hebron, along with the road from al-Auja to Beersheba through what had been designated as part of the proposed Jewish state. This would represent, with minor changes, the highwater mark of the Egyptian advance.

In the centre and south, Arab armies had retaken most of those areas allocated to a proposed Palestinian Arab state, but which had been lost to Zionist attacks before the declaration of the State of Israel. Despite making negligible advances into the UN designated Jewish state, they had conquered 13 Jewish settlements, only one of these was within the Partition Plan's proposed Jewish state, plus the Jewish quarter of Old Jerusalem.

On the other side, the Israelis welcomed the UN Truce because they needed time to build up and train their forces. Furthermore, the fact that the substantial settlement of Negba had not fallen meant that the Egyptians were unlikely to advance farther. For its part, Israel had captured 14 Arab towns or smaller townships along with 201 out of a total of 219 Arab villages within territory that the Partition Plan allocated to the Jewish state, plus a further 112 villages within the proposed Arab state. In purely military terms, Israel's strongly fortified settlements had proved to be powerful citadels that the Egyptian and other Arab armies had found very difficult to overcome.

GROUND OPERATIONS

As dawn broke, the Israeli assault force attempting to withdraw from ground outside the Iraq Suwaiden police fort was still pinned down by machine gun fire from the corners of the building. Suffering increasing casualties as the light improved, they were then hit by REAF aircraft from al-Arish. Following the Israeli attacks, and in particular the occupation of Hill 113, part of the Egyptian garrison in Iraq Suwaidan established another position astride the 'middle road' just south of the junction, but not too close to currently Israeli-held Kawkaba. This village, having been seized by Israeli forces late in May, would then change hands several times. Thus, the Arabs could still travel east and west whereas the Israelis could not travel north and south.

Although the Egyptian Army felt it had done well during the First Phase of Fighting – perhaps better than it had expected – there was no denying that the Egyptian advance had lost momentum. Communications were over-extended, staff-work often uncoordinated and ammunition stocks dangerously low. Meanwhile some equipment was proving faulty. On the other hand, morale was high, despite the fact that officers and men realised that the war would be a drawn-out affair rather than the speedy triumph their rulers had promised.

Both men and officers now needed rest and resupply, and some seem to have been sent to Egypt on leave even before the fighting ended. One such was Salah Badr, whose diary noted that he had arrived at a transit camp outside Gaza at 0400 hours. There he met some front-line colleagues, including another officer also named Salah, and together they remained in the camp until 1130 hours, awaiting transport to carry them on leave.

News of the Truce also reached the camp where Abd al-Rahman Inan and his REAF colleagues were held prisoner. Rather to Inan's surprise, one of the prison guards whom he knew as Moshe the Hungarian, was overjoyed at the news, telling Inan that the downed Egyptian pilots would be able to go home to Cairo 'within a week or so' while he, Moshe, could return to Hungary. Moshe apparently danced around the cell and spat on his own uniform, announcing that he would leave 'this country where there is neither work nor life', would see his family and old friends and would no longer have to learn Hebrew. As Inan pointed out in his book, Moshe the Hungarian was a bit too optimistic and did not return next day, despite Inan having promised to give him his own last reserve of cigarettes.

On the northern front, the reorganised Syrians in Mishmar HaYarden resumed their advance towards the main road on 11 June. However, they were then blocked by Oded Brigade units and newly arrived Carmeli reinforcements. By then the two sides had agreed upon the UN Truce, and the situation on the ground stalled until a ceasefire came in effect. The Arab Liberation Army (ALA) was still active in Galilee, and while Lebanese troops reorganised themselves and consolidated their current positions, an ALA force led by Fawzi al-Quwaqji pushed into central Galilee. By 11 June its leading troops entered Nazareth, just before the First Truce came into force.

AIR OPERATIONS

However urgent the threat to the police fort at Iraq Suwaidan, the men at al-Arish could not do anything until dawn. Then, according to Egyptian Communiqué No.31, the arrival of daylight considerably simplified operations:

> At 0430 our bombers raided a number of armoured cars carrying troops towards Iraq Suwaidan and strafed them with heavy machine-gun fire. They dispersed in all directions but our aircraft pursued them and wiped them out. At 0600 our bombers attacked large concentrations of enemy armoured cars and troops which were brought up hastily to reinforce the defeated enemy in Iraq Suwaidan. The enemy suffered considerable losses in men and armoured cars, some of which attempted to flee in the direction of Negba, but they were strafed by our aircraft which then attacked the colony of Negba ... There was full cooperation between our artillery and Air Force in this operation. At 0700 our dive bombers strafed a large force of enemy armoured cars and lorries, full of troops, moving in the direction of Iraq Suwaidan, and wiped them out completely in this area. Then our dive bombers carried out another attack on Negba and Gat, considerably damaging fortifications with high explosives and incendiaries. From all these operations our aircraft returned safely to their bases at exactly 0755.

This was clearly one of the REAF's most successful days during the Palestine War, and would be proclaimed as such across the Egyptian media. It also featured prominently in Major Muhammad Kamil Siddiq's article, 'Air War in Palestine' 'I do not exaggerate if I attribute the success of this battle to the air force'. Of course the al-Arish Operational Diary provides more detail. The four Spitfires which carried out Sortie No.1 in support of Iraq Suwaidan all took off at 0430 hours, al-Qarmuti in Spitfire 5 number 607, Wahdan in Spitfire 5 number 615, al-Abady in Spitfire 9 number 643, and Wafa' in Spitfire 9 number 633. Visibility was good, no enemy was reported, and according to the remarks column: 'The enemy (were) attacking our ground troops in Iraq Suwaidan and the latter told us to attack the enemy, so our planes attacked the enemy and succeeded in destroying three armoured vehicles. The enemy then retreated. Note that our planes were in (direct) radio contact with our ground troops all the time to help guide the attack (by the aircraft)'. All returned safely at 0610 hours.

This sortie was followed up by a second by four Spitfires 9s carrying high explosive and incendiary bombs that took off at 0600 hours or 0610 hours. According to the Operational Diary, one of the aircraft, Spitfire 9 number 643, had already taken part in Sortie No.1, which is unlikely. This time, Mustafa al-Utaify was at the controls and in all likelihood the serial numbers recorded in the Operational Diary got muddled in the urgency of the situation. Al-Utaify was reportedly accompanied by al-Abady, who had also taken part in the previous attack and now flew Spitfire 9 number 646. The other two pilots were al-Tarazi in number 651 and al-Wasaimy in number 657. Visibility remained good and there was no hostile fire, the comments column stating that:

> Our planes attacked enemy ground troops and their armoured vehicles which were heading for the battle to rescue the previous troops who had incurred heavy losses in equipment and lives. There was close cooperation between our air force and ground troops. Our artillery carefully identified targets with smoke bombs and we hit and destroyed a truck heading for Negba. After destroying the enemy, our aircraft attacked with bombs and strafing, after which white smoke came from an ammunition store.

Sortie No.3 again targeted Iraq Suwaidan and Negba in support of Egyptian troops, but this time only two aircraft were sent, probably because al-Arish was working at maximum capacity. No information was provided about their weapons loads, but Spitfire 9 number 633 was flown by al-Qarmuti, and Spitfire 5 number 607 by Wahdan. Both had taken part in Sortie No.1, though in different machines. They took off at 0725 hours, some ten minutes before Sortie No.3 returned; they themselves returning at 0820 hours to report good visibility and no hostile fire. According to the comments column: 'Our planes attacked enemy lorries full of soldiers which were south of Iraq Suwaidan who were trying to escape to the Negba settlement'.

If these Israeli troops were indeed south of Iraq Suwaidan police fort or Iraq Suwaidan village, they would have had to cross the Egyptian-held lateral road to find safety in Negba. The comments column continued: 'Also, two lorries were attacked and destroyed which were coming from (the direction of Egyptian-held) Falujah to Iraq Suwaidan. Then our planes were directed to the Negba settlement where they dropped their remaining bombs'.

Sortie No.4, as written in the Operational Diary, was a summary of the day's activities rather than a fourth mission, all the columns remaining blank except for the date. The time given for the first sortie was 0400 hours, which was probably when it was prepared rather than when it took off. The comments column simply reported that: 'Our planes undertook armed protection for the ground troops and returned at 0830'. Beneath this final entry for the First Phase of the Palestine War some concluding remarks were added, the first stating that: 'At 8:00 (0800 hours) firing has been suspended as ordered by Zamzam (Egyptian HQ radio) wireless station and all our planes returned to their shelters safely'. A General Note or second comment recorded that: 'After landing safely all the pilots and soldiers cheered His Majesty, Air Marshal, and our Commander in Chief (King Faruq)', this is followed by the unfortunately illegible signatures of a *Qa'id Liwa* (Group Captain) and a *Qa'id Asrab* (Squadron Leader). The Group Captain was probably signing off the entire section of the Operational Diary concerned with the First Phase of the Palestine War.

Although the al-Arish Operational Diary is the most important document to have become available for official Egyptian sources, others also shed light on the REAF activities during the First Phase of the Palestine War. Thus, we now know that there were sixteen REAF strikes aimed at Tel Aviv harbour and its facilities during this period, and that one or more Egyptian Lysanders had carried out two reconnaissance sorties on 15 May. Israeli claims that REAF attacks during May had only 'nuisance value', causing minor casualties and were largely psychological in their impact, were exaggerated if not downright untrue. Nevertheless, it is remains clear that against the sophisticated entrenchments and numerous well-supplied, well-armed garrisons of the targeted Israeli settlements, the small bombs and cannon-fire of Egyptian Spitfires had limited effect.

The Egyptians, like others before them, also discovered how vulnerable Spitfires were to ground fire. However, the only major setback suffered by the REAF during this First Phase was its disastrously mistaken attack on RAF-occupied Ramat David. Otherwise, the Egyptians' inability to prevent the creation of the Israeli Air Force's first real fighter squadron, equipped with potent though difficult to fly Avia S-199s, would cause serious problems in the future. In other respects, the REAF succeeded in its primary roles of direct support to Egyptian ground forces and reconnaissance. Egyptian aircraft had attacked Israeli Army convoys on their way to various fortified outposts and had struck at the strongly defended airfield outside Tel Aviv, as well as the vital harbour installations. Egyptian aircraft had similarly supported the Transjordanian Arab Legion, maintained Arab air superiority at least during daylight hours, and had struck various strategic targets.

The value of publicity was also understood by the REAF, as indicated by the list of targets attacked and losses inflicted painted

Through mid-May 1948, the REAF regularly attacked Tel Aviv. These two buses were destroyed at the central bus station. (Albert Grandolini Collection)

on the side of a Spitfire 9 flown by Squadron Leader Muhammad Abu Zaid for the benefit of Egyptian news cameramen and reporters who visited al-Arish early during the First Truce. It is also unlikely that these successes were those of Abu Zaid individually.

A REAF list of its strength at the time the First UN Truce came into effect overstated Egyptian losses because one pilot believed to be dead was, in fact, a prisoner-of-war. This document, nevertheless, stated that 5 aircrew had been killed (in reality 4), 3 injured and 2 captured (in reality 3). The REAF's current strength stood at 40 aircraft, 10 having been lost including those shot down and those damaged and currently unserviceable. This would appear to include 2 Spitfires lost before the Ramat David incident, the 5 downed by the British around Ramat David, 1 Dakota damaged beyond repair, and 2 other unidentified machines.

According to Major Muhammad Kamil Siddiq, the REAF's transport squadron (No.3) from which the Dakota 'bombers' had been drawn, carried a wide variety of military equipment and spare parts from Cairo to the Tactical Air Force base at al-Arish. They also carried officers, soldiers, many wounded men, mail and even entertainment equipment to maintain morale at the front. Enemy aerodromes at Sde Dov and Nir Am had been attacked frequently, with Aqir being attacked less often.

Farther north, when the Truce came into force, Latrun was still in the hands of the Arab Legion. However, by constructing an alternative road, Israeli forces relieved pressure on Jewish New Jerusalem and broke what had threatened to become a full-scale siege.

In the hours remaining before the First Truce, the SyAAF was again in action and after reporting significant fighting on the ground, Syrian Communiqué No.29 reported that: 'Our Air Force gave effective and successful support to our ground forces until 0800. They raided the colonies of Rosh Pina, Ilanya (unidentified) and Najmat al-Subh (Ayelet HaShahar)'. According to Israeli sources the first of these raids mistook its target and instead bombed the police station in Tiberias. Similarly, the last SyAAF operation of the day, and of this First Phase of the Palestine War, attacked the Israeli settlement of Beit Yosef for the first time. This lay 15km south of Lake Tiberius and was struck with 50kg bombs, which reportedly caused no damage.

During the Ramat David incident, on 15 May 1948, the REAF Spitfires attacked and hit on the ground a number of RAF aircraft, including these two Spitfires. In turn, British interceptors then shot down a number of Egyptian Spitfires. (Albert Grandolini collection)

Also hit during the Ramat David incident was this RAF Dakota, which completely burned out. Its wreckage was photographed on 22 May 1948. (Albert Grandolini collection)

2

THE FIRST TRUCE (11 JUNE TO 7 JULY 1948)

Following cajoling and threats from Britain and Transjordan, the Arab nations agreed to the proposed ceasefire. Israel's more enthusiastic engagement in the truce had a secondary purpose, allowing them to exploit the situation and import weaponry and men to bolster their armed forces. With the First Truce in place, both sides now acted to gain support for their positions in the UN, whilst Count Bernadotte and his team of UN Observers policed the truce as best they could.

ASSESSMENTS: THE ARMS EMBARGO

To briefly review, the absence of domestic arms production, the predominance of British influence and supply, and the lack of contacts beyond Britain and certainly behind the 'Iron Curtain' significantly affected the Arab nations' ability to rearm and recruit. Beyond Egypt's purchase of Italian made fighter aircraft, the embargo period was poorly navigated by the Arab states. Further, the ceasefire highlighted their inadequate numbers of reservists. The Arab armies could reshuffle but not reinforce, leading to the meeting between the Arab Chiefs of Staff and Abd al-Rahman Hassan Azzam Pasha, in which they explained to him the parlous position of the military that required consideration before any resumption of fighting, prompting Pasha's response that they 'have to resume fighting if we are attacked'.

In stark contrast, Israel was able to largely bypass the embargo with supplies arriving from the United States and elsewhere.

A REAF Curtiss C-46 Commando with Egyptian military officers, civilian officials and a representative from a UN Truce Observation team. (Hinnawi archive)

Haidar Pasha, Egyptian Minister of Defence, as seen in a chat with REAF officers, in front of a hangar, probably at al-Arish. Notable are inscriptions 'no smoking' in Arabic, on the walls behind them. (David Nicolle collection)

ASSESSMENTS: EGYPTIAN FORCES

At the end of the First Truce on 7 July 1948, Arab lines of communication in Palestine, especially for Iraqi and Egyptian troops, remained dangerously extended. However, the strength of the Egyptian Army had risen from 15,000 to 18,000 men and would prove sufficient to repulse Israeli attacks during the Second Phase, though not as easily as Egyptian sources claimed.

During the First Phase, the REAF had effectively operated on two fronts, supporting the main Egyptian Army on the coastal plain, as well as the Egyptian Light Force and Transjordanian Arab Legion in what became the southern part of the Palestinian West Bank. Unfortunately, the significance of the support given by the REAF to the Arab Legion has never been properly recognised. The effectiveness and importance of REAF aerial reconnaissance was certainly appreciated within Egypt, but has again been largely overlooked by outside observers.

To take one example: an aerial photograph of the settlement of Dayr Sunayd (Yad Mordechai) and a corresponding interpretive drawing was published in the article by Major Muhammad Kamil Siddiq in the *REAF Magazine* of October 1948. It showed this settlement's elaborate defensive system, typical of so many Israeli settlements, with concrete pillboxes having multiple firing apertures, other firing positions, shelters and communication trenches. It also highlighted the elevated topography of the location, showing that the site had been chosen with defence in mind while noting that this fortress – for such it was – overlooked the main coastal road between Gaza and al-Majdal. Of particular interest to REAF pilots who had to attack Yad Mordechai was a concrete-reinforced entrenchment below the elevated water tower, interpreted as a possible firing position against aircraft attempting to demolish the water tower itself.

During the first four days of the First Phase of the war, the REAF had largely concentrated on gaining air superiority by destroying Israeli aircraft wherever they could be found. REAF Spitfires raided Nir Am and its airstrip three times. On the 16 May, the ex-RAF aerodrome at Aqir, south of Tel Aviv, was attacked, a number abandoned ex-RAF aeroplanes being bombed. Nearby, a grounded Israel C-46 Commando (ex-NX67923) was sufficiently damaged to remain out of service for a long time. Meanwhile, the REAF had to provide fire support and air reconnaissance for the Egyptian Army on the front lines.

The following few days, from 19 to 22 May, the REAF had turned its attention to the attempted destruction of Tel Aviv's port facilities, along with the oil refinery, military production sites and transport systems. At the same time its direct support of the Army involved the REAF in attacks upon Israeli settlements and in continuing air reconnaissance. Furthermore, the Egyptians were already concerned about the Israelis acquiring more effective aircraft, so REAF pilots not only searched for the enemy's Avia fighters, but also bombed or strafed ex-RAF aircraft, abandoned or simply scrap, where such material looked as if it might be useful to their enemy, even as a source of spare parts.

The Israeli Air Force had originally intended to organise its first proper fighter squadron at Aqir, but now decided to do so at Herzliyya, which was farther away from al-Arish. Egyptian records show that in addition to these targets, the REAF attacked Negba, Yad Mordechai, Kfar Sirkin airfield (previously called RAF Petah Tikva), though the Israelis were not yet using it, and again Sde Dov aerodrome. REAF strikes on the airstrip at Nir Am proved more immediately successful, obliging the Israeli Negev Squadron to move its aeroplanes to the better protected Dorot, 6km farther east. At 1930 hours GMT on 16 May, Cairo Radio had mentioned propaganda leaflets for the first time, announcing that: 'Our aircraft also dropped leaflets on the Jewish colonies asking them to surrender'. Israeli air historian Meiri Ilan noted that some elderly Israelis who took fought during this period, recalled Egyptian Westland Lysanders doing precisely this.

In his memoirs, Israeli Prime Minister Ben-Gurion admitted that the REAF had air control throughout the first period of the 'official' war, and that the REAF executed approximately 116 air strikes during the first phase of what Ben Gurion regarded as the Israeli War of Independence. These he estimated to have involved around 232 sorties. For their part, Egyptian reports claimed that ships had been attacked, and in some cases sunk, in or close to Tel Aviv harbour, along with damage to three small Israeli naval vessels that were attempting to interrupt Egyptian naval traffic between al-Arish and Gaza, while attacks on Aqir air-base reportedly destroyed a large transport aeroplane and two enemy fighters.

REAF personnel during a prayer, at al-Arish Air Base. (Abu Zaid collection)

ASSESSMENTS: IRAQI AND SYRIAN FORCES

Despite having suffered a few losses, the RlrAF remained generally confident at the end of the First Phase of the Palestine War. Its primary role had been to support the Iraqi Army contingent, formed of an infantry regiment and an armoured battalion in what would become the northern part of the West Bank. Transjordan as yet had no air force, so relied upon the Iraqis for air support around Jerusalem and the REAF south of it. The RIrAF's zone of operations was in practice somewhat blurred, ranging from the southern outskirts of Jerusalem to the southern end of Lake Tiberius, and in both extremities overlapping the Egyptian and Syrian air forces' operational zones.

The RIrAF currently had a manpower strength of 101 officers, of whom 95 were aircrew, and 1,066 other ranks, though only part of this force was initially committed to the struggle for Palestine. During the First Truce the Iraqis would reportedly send all currently operational aircraft to Transjordan, this reaching a total of 4 Gladiators, 6 Ansons and few light aircraft. The Ansons of both No.7 and No.1 Squadrons may have been commanded by Lieutenant Colonel Muhammed Ra'uf Hasan, and had been the first Iraqi aircraft committed to action. Quite when the next flight of Ansons arrived at Mafraq remains unclear, but they might have been sent to replace the first group when the latter were withdrawn to Iraq to begin re-equipping with Hawker Furies. During the first phase of fighting, a couple of Iraqi-registered Miles Aerovans belonging to the Royal Iraqi Aero Society, civil registrations YI-ABV and YI-ABW, had been sent to Transjordan to serve as air ambulances, but are said to have proved unsuitable.

According to some sources, a Flight of Iraqi Hawker Furies would arrive in Mafraq, either late in the First Truce on 6 July, or during the Second Phase of Fighting on 16 July. Given the major retraining involved in converting Anson bomber pilots to Fury fighter pilots, the latter date seems more likely. Whichever the case, these Furies, their pilots and ground support personnel are said to have been sent to Damascus shortly afterwards. According to Israeli sources, the RIrAF eventually committed 5 Hawker Furies and 4 Harvards to the struggle. The authors can find no evidence for Iraqi Harvards entering combat, though it is possible they were handed over to the Syrian Air Force. The Israelis seem to have believed that the Syrian Air Force lost two Harvards, presumably badly damaged rather than shot down as a result of Israeli ground fire, during May, but the evidence for this remains unclear.

ASSESSMENTS: TRANSJORDANIAN FORCES

A number of Arab Legion troops had become involved in the Palestine War even before Transjordan's official entry into the conflict on 15 May. But in April 1948, when King Abdullah decided to intervene in Palestine, he had no intention of committing the Arab Legion to fighting Israeli forces. His only concern, as agreed between Tawfiq Abul Huda Pasha and Ernest Bevin, was to occupy the Arab part of partitioned Palestine as a step towards merging that area with his Kingdom. However, the smooth occupation that Glubb Pasha and King Abdullah had hoped for turned into severe fighting.

A pilot of the Royal Iraqi Air Force, with one of the brand-new Fury FB.Mk 11 fighter-bombers, delivered to Nos.1 and 7 Squadrons in 1948. The aircraft next to him wore the serial number 248, while the one visible in the rear was 249. (David Nicolle collection)

ASSESSMENTS: ISRAEL

On the other side, the overall position of Israel remained uncertain. Its forces were under pressure and had suffered several defeats. Furthermore, Jewish Jerusalem was under siege and the Negev was largely cut off. Although Israeli forces were growing much more rapidly in numbers and firepower than those of the Arabs, they were in desperate need of time to complete their reorganisation and absorb new equipment.

On 18 June, a week after the First Truce came into effect, David Ben Gurion in his role as Minister of Defence rather than as Prime Minister, held a meeting attended by directors of the branches of the Israeli General Staff and brigade commanders. They discussed the situation and came to the conclusion that the First Truce came none too soon for Israel's armed forces. Moshe Carmel, one of the brigade commanders, summarised the situation concisely by saying that the truce 'came to us as dew from heaven'. In their reports, the brigade commanders asserted that: 'the formations were weary and worn out. Casualty figures, particularly in infantry battalions, were extremely high. It was vital that the men should be allowed to rest and recover and those reinforcements should be allocated to the Battalions'.

According to the Israeli historian Netanel Lorch:

> The brigade commanders described the burden imposed upon their formations and demanded the establishment of further garrison battalions to hold on(to) static defence positions, whereby a major portion of their forces was pinned down. Thus, they can move for offensive actions, and make use of the truce period for training and reorganisation. It was also suggested that area commands be established separately from the command over field formations, so that the latter could, whenever necessary, be moved from one area to another without undue difficulty.

After hearing these reports and suggestions from his staff and brigade commanders, Ben Gurion laid out his own plan for the next phase. In it he asserted that, with a renewal of fighting, the war would enter its decisive phase. From the Israeli point of view, the triangle of Tel Aviv, Haifa and Jerusalem was the most important area. The fall of just two of these three centres would mean defeat. Accordingly, he argued that Israeli forces should concentrate their efforts on the central front, to defeat the Arab Legion and break out of this triangle. The Syrian and the Lebanese forces could then be expelled. Once this had been done, the Egyptian army would remain alone on the battlefield. Ben-Gurion therefore specified Israeli goals and priorities against the various Arab fronts. To achieve them, intensified efforts by the political leadership and military high commands should be carried out on three tracks to increase the military capability of Israeli forces, as follows:

A Absorb the arms and war equipment that had flowed, and was continuing to flow, into Israel following the end of the British Mandate.

B Conclude new arms deals to achieve superiority in arms, especially in aircraft and heavy weapons.

C Continue the reorganisation of Israeli forces having merged the Haganah and the Irgun tzva'i l'eumi (Etzel) into a single military organization under one command as the Israel Defence Forces (IDF), which had officially come into being on 28 May 1948.

The Haganah had, in fact, tentatively secured its communications with the south by seizing control of the 'middle road' from Julis to Burair, plus the intervening villages of Kawkaba and al-Hulaiqat. The inhabitants of the latter had been expelled and mostly sought refuge in nearby Falujah.

ACTIVITIES DURING THE FIRST TRUCE: EGYPT

During the First Truce, the Egyptian Expeditionary Force was strengthened by the arrival of volunteers from the Sudan and North Africa, plus an enlarged Saudi Arabian Army contingent, while more 25-pounder field guns, 4.2-inch mortars and Spitfires were committed. Units were also moved. The enlarged 6th Battalion, which had previously been involved in the Isdud sector, was sent to the Bayt Jibrin and Falujah sectors. However, one of its senior officers, Gamal Abd al-Nasser, did not accompany it, having been shot in the chest near Isdud, shortly after the First Truce had supposedly come into effect.

Previously available, unofficial sources had suggested that the REAF's operational aircraft now consisted of around 20 Spitfire 9s, only half of which were serviceable, and a dozen Spitfire 5s, the majority of which the Egyptians had kept serviceable. These were supported by less than 25 transport and communications aircraft, the great majority of which were serviceable. Meanwhile, other aircraft were understood to have been kept in store because Egypt did not have sufficient pilots to fly them. Now that official records have been made available, more is known about the true state of the Egyptian Air Force at the start of the First Truce. For example, the REAF's operational strength (excluding training aircraft) on 11 June can be compared with its strength at the start of the conflict on the night of 14-15 May 1948:

Total number of aircraft: x 48 on 14-15 May; x 46 on 11 June.
Fury: x 0 on 14-15 May; x 1 on 11 June.
Spitfire 9 fighter/fighter-bomber: x 20 on 14-15 May; x 18 on 11 June.
Spitfire 5 fighter/fighter-bomber: x 8 on 14-15 May; x 9 on 11 June.
Lysander recce: x 2 on 14-15 May; x 2 on 11 June.
Hurricane met-flight: x 4 on 14-15 May; x 3 on 11 June.
Dakota transport-bomber: x 8 on 14-15 May; x 7 on 11 June.
Dove transport: x 4 on 14-15 May; x 4 on 11 June.
Anson transport-bomber: x 2 on 14-15 May; x 0 on 11 June.
Commando transport: x 0 on 14-15 May; x 2 on 11 June.

Thus, in overall terms the REAF was essentially the same at the end of the first phase of fighting as it had been at the start, perhaps because increased motivation resulted in improved serviceability, which more than compensated for the losses suffered. Egyptian losses in Spitfires had been replaced by previously unserviceable aircraft. The Dakota which had been lost had not yet been replaced, but two recently renovated Commandos more than compensated for this where transport operations were concerned. The Ansons had been moved from an operational role to become navigational trainers.

During the course of the First Truce, the REAF restocked its forward base at al-Arish with munitions, other supplies and, of course, recently overhauled aircraft. Most notable were the probably additional Spitfire 5s from No.6 Squadron, along with their pilots and presumably also ground support personnel. Some – perhaps all – the Spitfire 9s were sent back to the better equipped aerodromes around Cairo for overhaul, although several undoubtedly then returned to al-Arish. Even so, some British and American experts claimed that the REAF would not be able to sustain further operations for more than two months, a confidential report by US Military Intelligence estimating that the 'combat effectiveness of the REAF was so low, it could not be considered an adequate defensive force'.

According to Lieutenant Ali Muhammad Mahbub's article, published a little over two months later:

> At the beginning of the first Truce period, the Headquarters of the Tactical Air Force received (the following) military instructions from the Commander-in-Chief:
>
> I intend to prevent the enemy from supplying its colonies located in the Egyptian occupation zone.
>
> This area is determined by the line … – … (censored).
>
> As long as we are in the truce period, you are allowed to strike only in two cases: (1) Self-defence, (2) To prevent the supply of enemy by land and air in the area of occupation referred to in item 2 … (Lieutenant Mahbub then requested his reader to note that the terms of the UN Truce were fully implemented).
>
> The means to implement the purpose are: (a) Strikes on any ground or aerial movement of the enemy entering the area specified by … [the limit of the Egyptian occupation zone] (here censored), (c) All hostile aircraft moving south of this line … – … (censored) are [to be] fired upon … [etc.] (censored).

In fact, the REAF had to respond to several Israeli breaches of the UN ceasefire, the most concerning of which were persistent smuggling of munitions through Egyptian lines to Israeli forces in the Negev. It seems that, as part of the effort to keep an eye on developments in the Negev, some of the REAF's reconnaissance missions were now longer-distance sorties from airfields close to Cairo.

REAF Headquarters was also painfully aware of the deficiencies that had become apparent during the first phase of fighting. These were listed as a low numbers of pilots trained in reconnaissance and bombing; a deficiency in technical instruments, such as bomb sights and radios; the poor quality of radio communication from al-Arish airfield, the poor quality or inadequacy of bombs used against strongly fortified bunkers; the low number of dedicated reconnaissance aircraft; the very low number of intelligence officers trained to interpret aerial photographs; the fact that the only airbase near the front was al-Arish which was not very suitable; the inadequate training of maintenance staff at al-Arish. After some aeroplanes came down in Israeli territory, the REAF also realised that it needed oxygen cylinders to enable pilots, particularly those in Spitfires, to fly at higher altitude.

In his report to HQ at the start of the First Truce, Wing Commander Sidqi Mahmud al-Milaigy suggested that al-Arish should have two fighter squadrons, each of 20 aircraft and, at a later date, a fighter bomber and a night-fighter squadron, again of 20 aeroplanes each. More realistic but also optimistic, Sidqi Mahmud urged that work should start at once on enlarging the al-Arish airbase, and providing more technical support along with a radar system. Furthermore, he suggested that forward airfields be constructed at the existing small airstrips outside Gaza and Falujah.

Meanwhile, outside Cairo, work continued on the commandeered Hawker Fury prototype, apparently at Hilwan rather than Almaza. An official complaint had been sent by the British Embassy to the Egyptian government on 14 June, stating that this aeroplane was used in action while still bearing British civilian registration markings. In reply the Egyptian Prime Minister said he would investigate the matter. An aide-memoire attached to the letter referred to aircraft in the plural, but, as no further mention was made of a second aeroplane, this was probably an error or misunderstanding. What is clear is that the Fury's British registration was removed during the First Truce. The machine was also given proper REAF markings and the serial number 701, but remained in an overall varnished metal finish.

Meanwhile *Qa'id Sirb* Muhammad Abd al-Hamid Abu Zaid returned from active service over Palestine. He advised on the Fury's new armament, while painting a skull and cross-bones on the Fury's spinner – a motif he had already added to at least one Spitfire. This machine's macabre decoration may have been a morale-boosting exercise or may have been in preparation for a role in a forthcoming feature film called *"A Girl from Palestine"*. In the event the hero of this film flew a different aircraft in his role as a REAF Squadron Leader. On the other hand, the Egyptian press did begin referring to Muhammad Abu Zaid as 'The Death's Head Squadron Leader'.

Poster announcing the film 'A Girl from Palestine'. (David Nicolle collection)

A fire-engine and crew of the REAF's own emergency service acting as extras in the film 'A Girl from Palestine', racing to 'help an injured Spitfire-pilot'. (David Nicolle collection)

Subsequent comments by the British Air Attaché, Air Commodore Hayes, were complementary regarding the REAF's engineering personnel during this period. He wrote that:

> … a great effort was made to repair their aircraft, particularly Spitfire aircraft, even to the extent of undertaking extensive repair work which even during the 'Battle of Britain' days would not have been carried out in the United Kingdom … Many Commando transport aircraft were overhauled, repaired and fitted for bombing (in fact only a small number of Commandos were made into bombers), and many Spitfire airframes and Merlin engines were modified locally to give increased speed and performance for these aircraft.

Major Muhammed Kamil Siddiq, in his article 'Air war in Palestine', was, as might be expected, even more complementary, claiming that; 'Thanks to the efforts of the air force workshops, the supply of planes, machinery, groups (sic.), radios and the like continued. All the necessary assistance to the army (was) as we have already mentioned'. The reality was not quite as good.

Considerable efforts continued to be made to provide the REAF with more bombers; these focusing upon the Transport Squadron, which was already working at full stretch carrying troops and equipment between Cairo and al-Arish. It was clear that the Egyptian Government had made an error when it did not purchase more of the supposedly scrap – or at least redundant – Dakota and Commando transport aircraft available at the ex-US air-base outside Cairo prior to the Palestine War. Those that were now part of the REAF's inventory were also suffering from a lack of spares.

Some of the simpler spare parts were reportedly manufactured locally and, with the support of civilian volunteers, REAF engineers had made some of the larger aircraft into what were referred to as 'strategic bombers'. In the case of the C-46 Commandos this entailed modifying existing equipment designed to drop supplies by parachute out of the side cargo doors, so that bombs could be rolled out instead. It was not particularly successful and was notably inaccurate. According to Munira Kafafi, the cargo planes' weight lifting capacity was also increased but unfortunately the measurements she provides in her book either do not make sense or are those of an unmodified C-46. She was, however, correct in recording that a lack of proper bomb-sights proved a major problem for the crews of all the REAF's so-called 'heavy bombers'.

Air Commodore Hayes' report went on to mention the REAF Engineering Department's attempts to make airworthy two American P-47 Thunderbolt fighters found in abandoned Allied scrap-heaps. There is no further evidence for this and it seems possible – if unlikely – that the British Air Attaché's sources mistook one or more Vultee Vengeance dive-bombers for P-47s.

REAF teams certainly scoured ex-RAF and USAAF military dumps in Egypt for anything that might be useful, especially for Spitfires, while German 250 and 500kg bombs were reportedly retrieved from abandoned Luftwaffe airfields in the Western Desert. These were renovated by REAF technicians, presumably so that they could be dropped from British-made bomb racks. Other Egyptian craftsmen produced locally manufactured incendiary bombs, almost certainly versions of the British 25lb (11.34kg) incendiaries mentioned in other sources. Consequently, the REAF was never short of such weapons.

Reference has already been made to civilian engineers enlisted by the REAF during the Palestine War. According to Air Commodore Hayes: 'To assist the service personnel in this work a force of approximately 3,000 civilian mechanics was enlisted by the Royal Egyptian Air Force. These mechanics had considerable experience having worked with the Royal Air Force and American Air Forces during the 1939/45 war'.

The training of new aircrew, especially pilots, continued throughout the First Truce, with increased intensity. One of those young men who were about to complete their advanced training was Wagdi Fadil. He was now with No.6 Squadron, at Hilwan and his logbook for this period makes interesting reading.

> 22nd June, Spitfire 5, number 619, solo, air test.
> 24th June, Dove, number 901, first pilot Mustafa Isma'il, second pilot self, fly to Almaza (a short hop from Hilwan).
> 24th June, Spitfire 9, number 623, solo, (return) to Hilwan.
> 29th June, Anson, number 516, first pilot Taha Abbas, (as passenger?) to Almaza (from Hilwan).
> Page of log-book signed off by REAF Hilwan Air Station, Sq. No. 6. (unnamed).

During the closing stages of the Second World War, the RAF re-armed several squadrons deployed in the Middle East with Spitfire LF.Mk 9s – like this example of No.208 Squadron. When replaced by advanced marks, the majority of these were either sold to Egypt and Greece or scrapped in situ. How carefully the scrapping was undertaken, and the quantities of spare parts the Egyptians were capable of recovering – remains unclear. (Albert Grandolini Collection)

In place of older marks, the RAF units introduced to service Griffon-powered – and much more powerful – Spitfire F.Mk XVIII, 22 or 24, like this example. The acquisition of such advanced aircraft was attempted multiple times, both by Egyptian and Syrian governments, but they remained outside their reach. (Albert Grandolini Collection)

Mustafa Shalabi al-Hinnawi was meanwhile in the earlier part of his advanced training at Almaza, mostly in Harvards. During June 1948, a month before he graduated, he and his fellow students had been shocked to find that their flying was severely interrupted by a lack of spare tail-wheels for the Harvards. As al-Hinnawi later wrote:

> This would delay our graduation to become pilot officers, and we had been waiting for this day for six academic years, which had never happened in any armed forces in the world!! So, I headed in my private car and under my initiative to Wakalit al-Balah (a junk yard where one could buy old or new items) and to Mr. Cassiss Musa Sulayman, the biggest junk dealer in Egypt, and asked him about small wheels for airplanes. He sent me to a large open plot land next to the Nile where I found a huge quantity of new tail wheels for Harvard airplanes as part of British Aviation scrap. So, I returned to him but did not tell him the real reason for my needing these wheels, and (instead) told him they were appropriate and that I would use them to make small (pedal) cars for children. So, he agreed after some haggling to sell me each wheel for 30 piasters, no more. I used 52 pounds of my private funds to buy these wheels.

Thus, the Harvards could start to fly once more.

There were also changes amongst the flying personnel at al-Arish. For example, during or perhaps even before the First Truce, Ali Sharmi seems be transferred to transport aircraft. In fact, Ali Sharmi is understood to have flown only one Spitfire mission during the Palestine War before being transferred. But, as this was before Operational Diary pages become available it cannot be confirmed. The official reason for his transfer is unknown, but Ali Sharmi's family regarded him as their 'black sheep', with a reputation for heavy drinking and womanising though also very daring – perhaps too much so. A later colleague amongst transport pilots maintained that Sharmi was one of very few Egyptian Air Force pilots with whom he would not have wanted to share a combat mission, either because of his unpredictability or his unreliability.

Training pilots and overhauling or virtually rebuilding aircraft was one thing; getting hold of new fighting machines was quite another. Fortunately for the REAF, Italy was keen to supply them. According to a British Foreign Office report, negotiations for the purchase of 24 Macchi MC 205 fighters were undertaken by His Excellency Abd al-Rahman Hakki Bey, the Minister Plenipotentiary in the Egyptian Legation in Rome; not Wing Commander Ibrahim

Mustafa Shallabi al-Hinnawi standing on the wing of a REAF T-6 Harvard serial number 415, at Almaza in the summer of 1948. (Hinnawi archive).

Three REAF fighter-pilots – Mostafa Wigdan Hafez (left), Thutmose Ghibrail or Ali Sharmi (centre), and Khalil el-Arousy – as seen at el-Arish sometime before 22 May 1948. During the truce lasting from 11 June to 7 July 1948, the REAF rotated out the second group of its pilots from the combat zone, and replaced them with another group. (David Nicolle collection)

Hakki. Other sources state that these negotiations were carried out through a commercial company called Egi-Italia, run by Abd al-Latif Ragailah, an Italian businessman of Egyptian origin, helped by Colonel Umar Sayf al-Din and 'flight officer Colonel Jizrin'. The latter was probably Ibrahim Hassan Gazerine, one of Egypt's early military pilots. The Egyptians' Aermacchi counterpart was retired General Eraldo Ilari, general Manager of the Company who had been Deputy Chief of Staff of the Italian Air Force after the fall of Fascism in 1943.

The first contract agreed between Aermacchi and the Egyptian government, officially designated E1, was signed on 23 June 1948. It was for 24 Macchi MC 205 Veltro fighters, 16 of which had originally been constructed as MC 202 Folgores. They would now be brought up to MC 205V standard through the use of more powerful Daimler-Benz DB 605 engines. According to a letter written to the author by the Italian aviation historian Giani Cattaneo, who worked for Macchi for more than 30 years, these MC 202s and MC 205s were Italian Air Force machines which had been standing almost derelict or were currently being overhauled at the Aermacchi factory. The cost to the Egyptians was not to be 'partly paid in cotton' as sometimes suggested, but would be the equivalent of £648,298 sterling – all to be paid in cash. Flight testing and acceptance was to be performed in Italy, with crating, transport and re-assembly in Egypt to be provided by an Aermacchi team as a separate service. The first two aircraft to be delivered, REAF serial numbers 1201 and 1203, were handed over before and on the day that Contract E1 was signed; namely 16 June and 23 June. They were followed by number 1202 on 2 July 1948.

Meanwhile, Fiat Aviazione was also keen to sell its aeroplanes to Egypt, and came in on the act shortly after Aermacchi. Thus, the REAF ordered a first consignment of 17 Fiat G 55A1 fighters and two G 55B fighter-trainers, with an option on 20 more advanced G 59s, though the latter were never purchased. As with the Aermacchi order, most of the G 55A1s had been retrieved from somewhat neglected stores in northern Italy and were then renovated in the Fiat factory. Others were newly built, while the G 55Bs were drawn from the Italian Air Force Flying School at Galatina. Deliveries of these Fiats would nevertheless be slower than their Macchi rivals, not starting until September 1948, and the last aircraft of this first batch did not reach Egypt until the following year.

Egypt also wanted radar. According to a British Foreign Office memorandum of July 1948, the British RAF had agreed in principle that the Egyptians could use three existing radar installations in the British-controlled Canal Zone as a 'loan' which would not be subject to the embargo. They would be operated by Egyptian personnel under close British supervision, and their primary task was to watch

A Macchi MC.205V as seen while being prepared for a demonstration flight in attendance to an unknown foreign mission, at Venegone airifield. At the time, the aircraft was painted in anti-corrosion (light) green overall. (David Nicolle collection)

The REAF Spitfire Mk V serial number 602 was the former JK195 of the RAF. During the Palestine War, it flew at least seven combat sorites, most likely eight, and may have done more during the early weeks, before the REAF Tactical Air Force (al-Arish) Operational Diary becomes available. Its known sorties range from 30 June to 21 October 1948. Apart from two reconnaissance sorties, they consisted of ground-attack missions with bombs and guns, with the exception of the last – which was a standing patrol to protect al-Arish, resulting in combat with Israeli fighters and the loss of Mukhtar Sa'aid. Notably, with exception of the two sorties by Yahya al-Aidarus, all the missions completed by 602 were flown by different pilots. (Artwork by Tom Cooper)

REAF Spitfire Mk V serial number 616 was another workhorse of the Palestine War. It flew at least 19 operational sorties, the first of which was recorded in the al-Arish Operational Diary on 9 June 1948 (making it likely additional, unrecorded missions were flown earlier). Indeed, 616 often flew multiple missions per day: notably, on 10 July, when it took part in Missions 4, 6, and 11. Its last recorded mission was between 1525 and 1600 hours of 18 July 1948: a ground attack against Israeli positions in the abandoned Palestinian village of al-Burair, flown by al-Qarmuti. Three hours later the UN Truce came into effect, which would last until October: during this truce, REAF withdrew its surviving Spitfire Vs from front-line operations – until a few were hurried to the battle, later on, but not 616. (artwork by Tom Cooper)

When King Abdullah of Transjordan visited Cairo on 22 June 1948, he travelled in a Douglas DC-3 of the BOAC. While taxiing to the disembarkation point, the aircraft had the flags of Egypt and Transjordan displayed above the cockpit. The BOAC crest and name were applied in dark blue on the front of the fuselage and below the cockpit, the writing being B*O*A*C Speedbird, with white bars behind the inscription B*O*A*C. Although the registration of the aircraft in question remains unknown, the BOAC had a DC-3 registered as G-AGNG at the time in the Middle East, which makes it likely that this was the aircraft on which King Abdullah flew. (Artwork by Luca Canossa)

The REAF Curtiss C-46 serial number 1002 was distinctive in retaining a rare camouflage pattern used by the USAAF in the mid-1940s. This included sandy brown on upper surfaces of the fuselage, olive drab on top wing surfaces, rear fuselage and the fin, and light grey on undersurfaces but extending quite high up the sides. US insignia was probably obscured by some olive green and the Egyptian national markings and serial applied instead. The serial 1002 suggests that this was one of the first Commandos acquired by the REAF. (Artwork by Luca Canossa)

As the C-46s were progressively put through overhauls at REAF depots – and often, virtually, re-built – they began receiving very different camouflage patterns. In the case of the aircraft with serial number 1003, this seems to have consisted of the olive drab used by the USAAF, with extensive sandy yellow patches that, though following a similar general pattern, varied from machine to machine. Often enough, these variants resulted from the requirement to cover different smaller US markings. By contrast, application of national markings and serial numbers always followed the same pattern. (Artwork by Luca Canossa)

The differences and similarities in camouflage patterns between individual REAF C-46 Commandos are obvious when comparing the two examples illustrated above, with the serial number 1025 shown here. Indeed, its relatively high serial number might be indicative that it did not enter REAF service until towards the end of the Palestine War, or had even been kept back as reserve. Sadly, serial numbers of the handful of REAF C-46 modified to carry bombs remain unknown. (Artwork by Luca Canossa)

Assigned to the REAF Transport Squadron, this particular Douglas C-47 Dakota – serial number 811 – was also modified for use as a bomber, and given an overall sandy brown camouflage pattern. It might have first been mentioned in the REAF Tactical Air Force Operational Diary on 18 July 1948 when, escorted by a single REAF Spitfire LF.Mk 9, it attacked the fortified Israeli settlement of Revavim, but this is uncertain. Serial number 811 was certainly used as a transport from al-Arish to Cairo on 3 November, after which it may have undergone significant modifications converting it into a night bomber. This saw the creation of a position for the bomb-aimer, apparently including the removal of fuselage panelling along the lower hull. Alternatively, a machine gun or a light cannon was installed instead. The aircraft flew at least one known sortie in this configuration: late on 27 December 1948, while crewed by six, it attacked the Aqir Air Base. The C-47 serial number 809 flew a similar sortie in the early hours of 30 December, also including an extra crewmember: a possible gunner. (Artwork by Luca Canossa)

The REAF C-47 Dakota serial number 816 is best-known for being photographed during a visit or stop-over in Cyprus, in around 1950. Obviously, it was not the only Egyptian C-47 to wear the same camouflage pattern in olive drab and sandy yellow on upper surface, like most of REAF C-46s, but it is one known to have flown at least one single bombing sortie against Israeli targets in 1948, which took place on 14 July. It is possible that at the time the aircraft was still in the high-speed silver overall finish. (Artwork by Luca Canossa)

The C-47 Dakota serial number 818 seems to have been used as one of the transports maintaining the vital air bridge between bases in the Cairo zone and the Egyptian expeditionary forces operating in southern Palestine. The aircraft survived the war and was still in service in the early 1950s, when it was last seen, still wearing its overall high speed silver finish, plus a small black anti-dazzle panel in front of the cockpit. (Artwork by Luca Canossa)

The Spitfire LF.Mk 9 with serial number 623 was one of the first acquired by REAF. This is a reconstruction of its look as of late spring and summer 1948, when it was already showing the signs of wear and tear, but had also received a sort of REAF roundel around the front of its spinner. The black Arabic serial number was obviously applied by hand and brush, with little uniformity. Between 16 July 1948 and 2 January 1949, serial number 623 is known to have flown at least 13 combat sorties: it survived the conflict and remained in service for at least a year longer. The inset shows a 50lb bomb (essentially a mortar bomb calibre 120mm), with Arabic inscription, 'Guide me to Tel Aviv' applied in white and black. (Artwork by Tom Cooper)

This is the same Spitfire LF.Mk 9 serial number 643, shown in the colour section of the Volume 10, but now with complete national markings, a partially repainted rear fuselage and fin, and an additional line with mission markings applied ahead of the cockpit. The line might have read, 'Egypt will be victorious', and was probably applied when the aircraft featured in the film 'A Girl from Palestine', made in summer 1948, when 643 was based at Almaza, rather than al-Arish. That said, the colours used remained the same like on all REAF Spitfires of the time, including BS381C/638 Dark Sea Grey and BS381C/641 Dark Green on top surfaces and sides: undersurfaces were painted either in BS381C/108 Aircraft Blue or an even darker colour, instead of the usual BS381C/697 Light Admiralty Grey (the latter was retained on the spinner only). (Artwork by Tom Cooper)

The Spitfire LF.Mk 9 serial number 648 appears to be one of the aircraft re-assembled from more than one derelict and abandoned British or US machines left behind in Egypt after the Second World War – as described by Air Commodore Hayes in his first Annual Report on the REAF after the Palestine War. As a result, it received a non-standard camouflage pattern, poorly applied in a hurry (without the first layer of clear varnish), and thus shows signs of wear, especially around the engine. The spinner was re-painted with dark green at the rear, dark grey in the middle, and Egyptian tricolour at the front. When featuring in 'A Girl from Palestine', the undersides appeared dark, but not black: Aircraft Blue or Azure Blue is most likely. Notable is the large underwing hardpoint with a 250lb bomb (125kg). The aircraft was probably held back at Almaza or Hilwan for advanced training purposes or for the defence of Cairo, and appeared in the REAF Tactical Air Force Operational Diary only in early July, and then in October 1948. (Artwork by Tom Cooper)

The Hawker Fury prototype had originally been known by its RAF serial number, NX798, but since 29 January 1948, it wore the British civilian registration G-AKRY. Its armament and under-wing stores were in place for display in England, but removed before it was flown via Malta to Heliopolis, on 24 April. Hawker test-pilot Bill Humble took the Fury up for a test flight on the 27th, followed by Qa'id Sirb Halim Zaki and, possibly, Sidqi Mahmud al-Milaigy. Both Egyptians returned full of enthusiasm, whereupon the Egyptian government impounded the airplane and moved it across Almaza aerodrome: from the BOAC airline hangar to the REAF hangar, and that while it was still wearing the livery shown here: high-speed silver overall finish, with a prominent cheat line in red, and the full civilian registration. (Artwork by Tom Cooper)

While Humble returned home to face the unfair accusation that he had allowed the Fury to be stolen, the British government protested to Cairo and lengthy negotiations followed. By the time both sides agreed to make its sale official, on 30 December, the aircraft had been lost in action. In the meanwhile, it had been handed over to Qa'id Liwa Salih Muhammad Salih, the REAF's Chief Equipment Officer (or Quatermaster General): his team adding four long-barrelled 20mm cannon, and universal bomb carriers, pilot armour, and a gun camera. The aircraft still wore the British civil markings when rushed to al-Arish at an unknown date in early June 1948 (as shown here), but only Squadron Leader Abu Zaid was allowed to fly it on operational missions. On 14 June, the British Embassy in Cairo issued an official complaint to Cairo, stating the Fury had been used while still in British civilian markings. The Egyptian Prime Minister replied that he would investigate. (Artwork by Tom Cooper)

The British registration and the red cheat line were certainly removed from the Fury prototype during the First UN Truce (11 June – 8 July 1948) – when the aircraft was withdrawn from al-Arish, probably to Hilwan, and had not only the national markings applied instead, but also the small black serial number 701. Contrary to unsubstantiated reports in the Israeli and Western press ever since, however, it did retain its high-speed silver overall finish. Squadron Leader Abu Zaid painted the spinner partially in black and then applied skull-and-crossbones in white: the same motif he had previously applied on the Spitfire Vc serial number 620. After its first recorded action, on 11 July 1948, Abu Zaid flew the serial number 701 into at least seven further combat sorties, on the last of which, on 19 October 1948, both the aircraft and the pilot were lost. (Artwork by Tom Cooper)

The RIrAF Anson serial number 186 is believed to be one of those deployed to Mafraq in Transjordan to support the Iraqi Army during the Palestine War. The two green stripes around the rear fuselage denoted either its assignment to a specific squadron, or to a specific flight leader in the unit. Otherwise, the aircraft retained its high-speed silver overall finish and the usual set of national markings (with triangles in six positions, and fin flashes). The serial was applied in Arabic numerals on the rear fuselage: it remains unclear if it was repeated on the nose, like on other Iraqi Ansons of 1948. (Artwork by Tom Cooper)

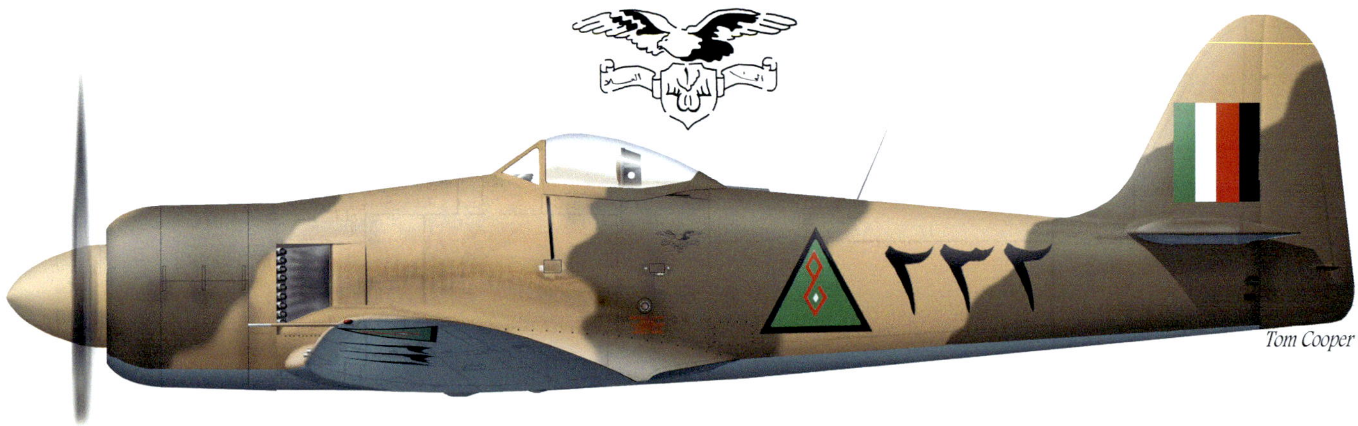

The Hawker Baghdad Fury serial number 232 belonged to the first batch delivered to Iraq, and is understood to have been one of a handful briefly deployed to Transjordan and Syria during the Palestine War. Applied before delivery, their camouflage patterns, national markings and serials were very consistent and included mid stone (BS381C/362) and dark earth (BS381C/350) on top surfaces and sides, and pale blue (BS381C/172) on undersurfaces. The only other markings accepted by the RIrAF were those of No.1 Squadron: an eagle and a scroll with Arabic lettering, applied in black below the rear section of the cockpit hood. (Artwork by Tom Cooper)

Early during the Palestine War, the SyAAF is understood to have maintained greater consistency regarding application of camouflage colours on its T-6s – compared with those delivered to Egypt. Correspondingly, all received different camouflage patterns in mid-stone (BS381C/362) and Brunswick Green (BS381C/226) like the serial number 205, show in the Volume 10. On 10 July, two SyAAF T-6s were attacked by a pair of Israeli S-199s: this was the serial number 206, piloted by Sergeant al-Abid, with First Mechanic Muhi al-Din Wadi as observer-gunner. During the ensuing exchange of fire, Wadi hit the Israeli fighter, which crashed outside Qunaitrah, killing Lionel Bloch, from South Africa. The T-6 returned to base, but Muhi al-Din Wadi subsequently died of injuries sustained in this engagement. Six days later, serial number 206 was either shot down or suffered catastrophic structural failure as a result of the previous combat and crashed at Ayelet HaShahar, killing the crew of Faisal Nassif and Ali Nash'at. (Artwork by Tom Cooper)

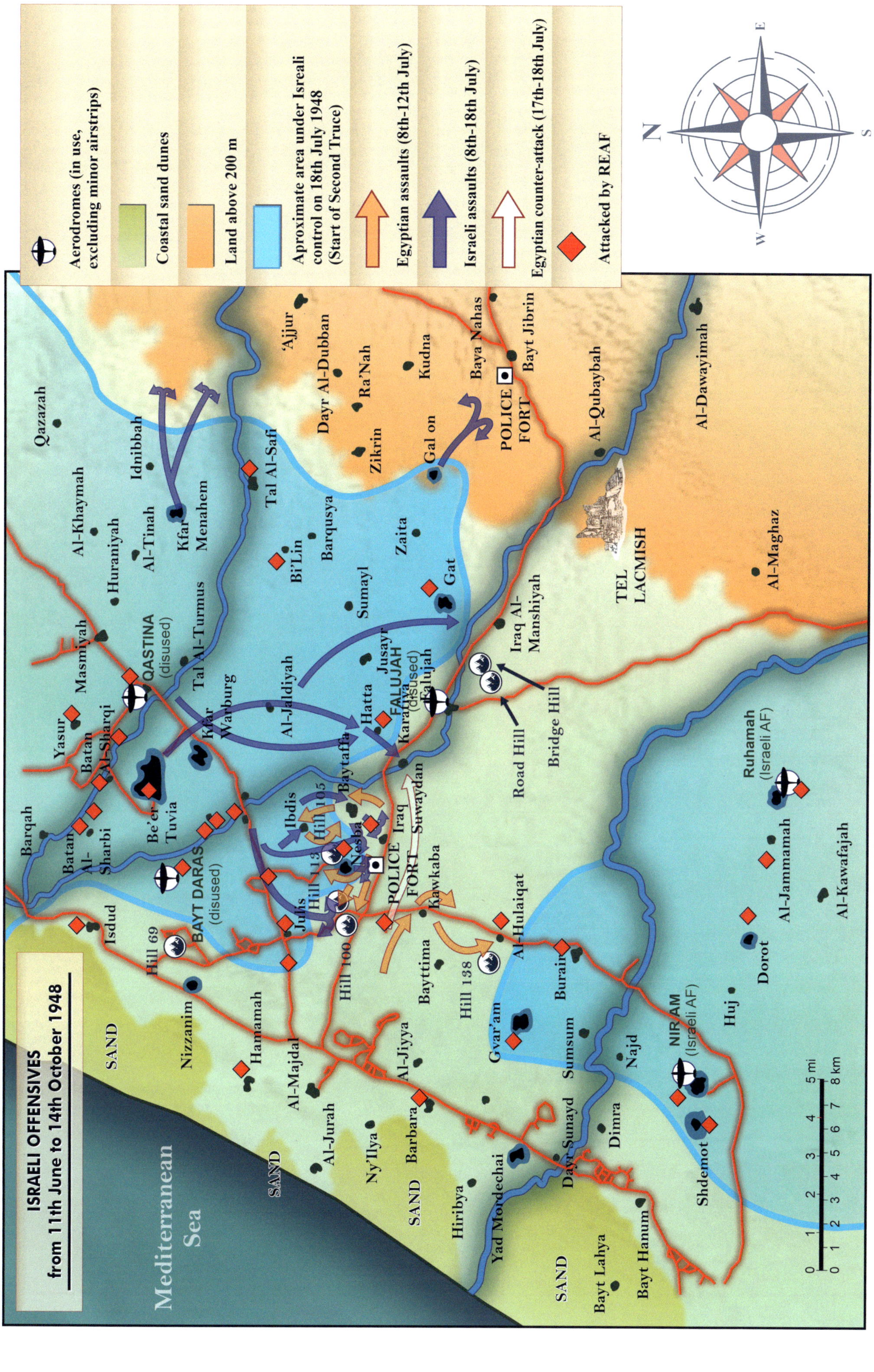

(Map by Anderson Subtil)

(Map by Anderson Subtil)

An overhauled Macchi MC.205V as seen in disassembled condition, during preparations for its shipment to Alexandria, at Valleo Olona, outside Varese, in September 1948. (David Nicolle collection)

A close-up view of the same MC.205. Notably, the aircraft was painted in Italian colour *nocciolo chiaro 4* (brown) overall. (David Nicolle collection)

Rear view of the same aircraft, as seen in the process of being packed in its transportation crate, at Valleo Olona, outside Varese. Notable is the right wing lying next to the fuselage. (David Nicolle collection)

The first Fiat G.55B two-seat conversion trainer assembled in Egypt. The aircraft subsequently received the serial number 1251. (David Nicolle collection)

for Israeli aircraft which might seek to penetrate Egyptian air-space, hiding amongst British aircraft operating in the area. In return the Egyptian government had agreed to relax some of the restrictions that had been imposed upon British night flying near the Canal Zone since the start of the 'official' Palestine War.

Meanwhile, three captured REAF pilots were still languishing in Israeli prisoner-of-war camps, dashing initial hopes that the UN Truce would result in their rapid release. According to Abd al-Rahman Inan, his cell overlooked a gorge and was closed by an iron door. It was in one of several buildings scattered across a camp which itself overlooked the sea. In the heat of summer these cells became intolerably hot, so that Inan would spend the entire day in his underwear rather than his REAF uniform.

Ten days into the First Ceasefire, a good-looking but rather sullen-faced young woman dressed in white brought Inan a package of clothing. When he asked if they were from the Red Cross, she told him they were from the prison administration. The package consisted of a military-style T-shirt, socks, handkerchief and underwear, which the prisoners apparently had to pay for. With nothing much else to think about, Inan's mind seemingly focussed on the ill-prepared, all too often bad smelling, and generally inadequate food prepared for the prisoners by an Armenian cook, who was himself a prisoner. Inan also fixated upon the erratic supply of cigarettes, recalling that during the first few months of his captivity he habitually smoked seven a day, often having to light the next cigarette from the former because prisoners were not allowed sulphurous matches.

Lack of exercise caused him further problems, as did his inability to tell the time because his watch had been taken away. As a practicing Muslim this meant that could only guess at the correct time for his prayers. Seemingly the only thing which cheered Inan and his fellow inmates was the wail of air-raid sirens in nearby Tel Aviv, heralding another raid by their REAF colleagues from al-Arish or Cairo. Sometimes they could even feel the vibration of the exploding bombs.

There were, of course, occasionally other diversions. One day, Inan and the Camp Commandant were sitting in a trench in the shade, discussing European classical music. At that moment, Barakat emerged from the toilet area limping and with a prison guard prodding his back with a gun. It was against the camp's regulations for prisoners to communicate but Inan was only ten metres away and saw what he described as the ghost of a smile cross Barakat's lips. The Commandant noticed this and gave the guard a public dressing down for allowing the Egyptians to get so close to each other. Soon every Israeli seemed to be joining in, each accusing the other of being at fault. This, of course, undermined the Commandant's status in front of his prisoners, so the discussion of music came to an abrupt end. Instead the Israeli officer began to tell Inan about the wicked things the British were supposedly doing to Egypt, such as 'selling the Sudan'. The promulgation of a new Constitution for Sudan on 19 June had, in fact, caused deep disquiet in Egypt.

Later, in June 1948, Inan was transferred to a different camp where he was put in a small concrete cell with a metal door and a hole in the ceiling for light. When he complained about this, he was transferred to another part of the same building which had an electric light. During his first night in this hell-hole he asked to be taken to the toilet. When this was refused, Inan heard a voice with a pronounced Egyptian accent telling him that he would have to 'pee in his cell'. It was a fellow prisoner, Gamal Irfan, who had been captured about a month earlier. In the morning, he heard female voices singing a popular song, '*Itdala'a Gamal Himdan*', so Inan altered the words and sang '*Itdala'a Gamal Irfan*". Once they realised that they were in neighbouring cells, they developed a code of knocking on the intervening wall to communicate, learning that neither were in good health.

On 7 July, the day before fighting resumed, the REAF prisoners-of-war were delighted to hear that a qualified chef – presumably a Muslim – would be allowed to cook the meals that would mark the breaking of their daily fasts during the rest of the Holy Month of Ramadan. Abd al-Rahman Inan recalled that these meals were remarkably good, given the very limited ingredients available, and included pancakes, sweets and concluding with cigarettes. Furthermore, the food which had previous been supplied to the prisoners meant that fasting during Ramadan was not so different to eating camp food at other times.

The impact of the first phase of the 'official' Palestine War was different within each country involved. In Egypt the conflict failed to unify the country and, although Egyptian intervention in Palestine reduced domestic opposition, it did not stop it entirely, even with the imposition of martial law. Egyptian communists were particularly targeted because they had followed the lead of the Soviet Union and had accepted the original UN Partition Plan. To some extent they almost came to be seen by the Egyptian authorities as pro-Zionist. Hundreds were therefore arrested or interned.

There was not only violence against some government institutions in Egypt, but also against the local Jewish community. By June 1948 around a thousand Egyptian Jews had been interned as suspects, while a few bombs had been placed in Jewish areas of Cairo. On 20 June, one such device destroyed four houses and killed twenty-one people. On 6 June, plastic explosive was found in a Sudanese agency in the Egyptian capital. On 9 July, another explosion killed 44 people and injured 55. There would be further blasts in September. This was accompanied by not particularly successful calls to boycott Jewish shops and businesses. The Muslim Brotherhood even encouraged attacks on 'the hat wearers', meaning the Orthodox Jews. Although the Egyptian authorities initially did little to combat such hostility to an Egyptian religious minority, it eventually decided that the Brotherhood was the major source of such tensions and banned the organisation. In response Egypt's Jewish community tried to show loyalty by subscribing to a fund in support of Egyptian troops at the front.

A conference of the Egyptian Army officers in command of forces deployed in Palestine, in late June 1948. (David Nicolle collection)

ACTIVITIES DURING THE FIRST TRUCE: IRAQ

Much less is known about the Iraqi and Syrian Air Forces during this period. RIrAF Avro Ansons were still based at Mafraq, in northern Transjordan, commanded by Lieutenant Colonel Muhammad Ra'uf Hasan. Two aircraft were lost during early July, seemingly without fatalities, but RIrAF morale is said to have remained high.

The idea that Iraqi air power was in decline was not, however, correct. According to Iraqi Air Force General Arif al-Razzaq, who was a young Lieutenant serving in No.7 Squadron RIrAF in 1948, when the Furies were delivered they were allocated to No.1 and No.7 Squadrons. Both these units were currently equipped with Ansons and were already deployed or partially deployed to Mafraq where they had been actively participating in the fighting. Now, the RIrAF took the whole of No.1 Squadron out of the battle so that its pilots could re-qualify on Furies. They were then sent back to Mafraq to reinforce No.7 Squadron, which was still under the command of Lieutenant Colonel Ra'uf Hassan. In what al-Razzaq described as 'the last morning before the truce' – apparently on 6 July 1948 – six Furies arrived at Mafraq. However, it is possible that there has been a misunderstanding or even a simple typing error here, and that these Furies arrived in Mafraq on 'the last morning before the truce ended'.

What is clear is that the RIrAF sent a small number of new Hawker Fury fighters to Transjordan. This transfer may also have been delayed because of the death of Fury pilot Captain Yahya al-Dabuni during his final training outside Baghdad. Following this loss, all Iraqi Fury pilots reportedly had to undergo very strict medical checks, as a result of which a significant number had been declared unfit for the physically demanding Furies. They were reportedly given just two months to get into shape. Ten of the 12 Furies ordered from Britain are believed to have reached Iraq by this time. Thus, the unit at Mafraq, still usually referred to as No.7 Squadron, was flying both archaic Ansons and very modern Furies. The fate or destination of the RIrAF Gloster Gladiators, which had been based at Mafraq, remains a mystery.

How long the Iraqi Furies remained at Mafraq, and whether they flew training or familiarisation sorties during the First Truce, is also unknown, though one was reportedly damaged in a landing accident at an unspecified date, the inexperienced pilot being unhurt. Six were then sent to defend the Syrian capital of Damascus, which had recently been bombed by the Israelis. This resulted in another Fury being damaged in a landing accident at al-Mizzah outside the Syrian capital, though again the pilot was uninjured.

ACTIVITIES DURING THE FIRST TRUCE: SYRIA AND TRANSJORDAN

The threat posed by Israeli bombers was clearly taken seriously by the Syrian Air Force. As a result, all transport, training and other light aircraft were evacuated to landing strips elsewhere, reportedly in the eastern part of the country. Only the SyAAF's armed Harvards and Ansons remained at al-Mizzah. Syrian Arab Airways had already been placed under military control around the start of the Palestine War, though the official decree authorising this take-over would not be signed until 12 May 1949, almost a year later.

The Kingdom of Transjordan did not of course have an air force, but the Arab Legion commandeered two Arab Airways Dragon Rapides, civil registrations TJ-AAC and TJ-AAP, which were then given the military numbers R300 and R301. However, the date of this is unclear and might not have been until 1949.

ACTIVITIES DURING THE FIRST TRUCE – ISRAEL

The notably successful efforts of the Israeli Air Force to obtain military and civilian aeroplanes from wherever this could be done – legal or otherwise – is well documented and well known.

In practice, the First UN Truce proved to be a turning point in the course of the war, with a clear shift in the balance of forces in favour of Israel. This Israeli achievement was possible not only because they had more experience at smuggling men and equipment than did the Arabs, but because they were supported by the Zionist Organizations' existing international facilities. To this should be added significant financial aid, notably from Europe and North America. The skilled use of such advantages resulted in an airlift of arms from Czechoslovakia, as well as Boeing B-17 Flying Fortress bombers and some fighters being flown in from the United States and Britain. The Israeli Army similarly obtained additional tanks, armoured vehicles and artillery. In the meantime, the Zionist Organisations intensified their efforts to encourage the immigration of young Jews, to make their *Aliyah* or 'ascent' from the Jewish diaspora to the new state of Israel. Zionist organisations similarly encouraged the continued recruitment of sympathetic volunteers with military skills and experience.

When it came to the creation of a truly effective Army, there were two schools of thought in Israeli at this time. One, led by Israel Galili, the Deputy Defence Minister, favoured a 'peoples' war' and an army designed for guerrilla operations. The other, led by Prime Minister Ben Gurion, favoured a conventional army. Ben Gurion had his way and Galili resigned.

There was no real manpower shortage in Israel, nor a shortage of light weapons. At a meeting of senior commanders at the Israeli GHQ in the middle of June, Moshe Zadok, Chief of the Manpower Wing at GHQ, announced that within days a further 3,000 boys of seventeen would enter training camp, while Yigal Yadin, the Chief of Operations, stated that he already had enough rifles for forty battalions, for second-line troops and for the home defence of all settlements. An orthodox Southern Front Corps Command was, in fact, currently being established to face the Egyptian threat.

While absorbing the men and war material, which arrived in a steady stream, the IDF continued its reorganisation and intensified training. In the process, the IDF High Command established four regional commands. Thus, by the end of the First Truce, Israel was able deploy a trained force of twelve mobile brigades. There were also additional garrison battalions; four brigades on the northern front, two brigades on the eastern front, three on the central front and three on the southern front.

Sadly, these activities were accompanied by the ongoing expulsion of Palestinian inhabitants of those areas under Israeli control. Most of this ethnic cleansing was reportedly carried out by Israel militias, rather than the emerging and undoubtedly more disciplined Israeli Army, though the latter were not entirely innocent of such crimes.

The number of expulsions of Palestinian Arab communities, and associated clearances and demolitions of Palestinian Arab villages, had decreased during the First Phase of the 'official' Palestine War and continued to decrease during the First Truce. Put bluntly, the various Zionist armed forces which evolved into the Israeli Army had other priorities – namely to defend the territory which they already held and to protect those Zionist, now Israeli, settlements which were at risk of being overrun by advancing Arab regular forces.

Nevertheless, the sources most critical of Israeli military behaviour during the Palestine War have listed twenty-two Israeli actions that they consider to have broken the laws of war. They reported four on the 20/21 May and one on each of the following days: 21st, the night of 22nd/23rd, 24th, 25th, 27th, 28th, and the night of 30/31 May. Others reportedly took place on the 7th, 10th, the night of 10th/11th and 16 June, while four more were listed as having taken place on unspecified days during that month, followed by two on unspecified dates in July.

As had been the case during the so-called 'civil war' period of the Palestine conflict, news of these events rapidly reached the neighbouring Arab countries and their armies in the field. Here they increased anger against the enemy, though fortunately most such forces were under sufficiently strict discipline that the troops' fury was rarely turned against prisoners-of-war or other captives. However, there were exceptions, usually where auxiliary or irregular units such as Muslim Brotherhood volunteers were concerned. Also, the small force of Saudi soldiers attached to the Egyptian Army would be accused of at least one atrocity.

The four weeks of the First Truce were undoubtedly vital for the IDF in general, and the IDF/AF in particular, giving the latter precious time for reorganisation and training. Reorganisation was particularly necessary because of a steady influx of aeroplanes, equipment and volunteers rich on combat experience from the Second World War. Meanwhile lessons learned during the First Phase of the conflict needed to be absorbed and digested.

Despite the fact that the UN Truce Resolution forbade the introduction of new military equipment and manpower into the area, both sides tried to bypass these restrictions while also using the time to improve their positions. It soon also became clear that the Israelis were disregarding the truce restrictions to a greater degree than the Arabs, and were doing so more effectively. Indeed, during June and July 1948, aeroplanes seemed to be flooding into Israel. In theory the UN arms embargo applied equally to all combatants, but in practice it hurt the Arabs while helping the Israelis. This stemmed from the fact that the Western powers, which were the Arabs' primary source of arms supply, observed the embargo whereas the Soviet bloc, who as yet had no relations with the Arabs, did not. Thus, during the four-week truce, the Israelis had a greater opportunity to improve their military position.

Lessons also had to be learned, especially in the light of the Arab air forces' greater than expected effectiveness. Some units were also moved. For example, following the REAF attack on Aqir aerodrome near Rehovot on 31 May, the IDF/AF's No.101 Squadron had moved its Avia S-199 fighters and its operational centre to Herzliyya landing ground. This lay east of Kfar Shmaryahu, north of Herzliyya town, and approximately 15km north of Tel Aviv. Here a 2,000 metre unpaved runway had been constructed, seemingly almost surrounded by orange groves, while a nearby modified water tower became the aerodrome's control tower. Equally significant was the transfer of the IAF's Central Maintenance Unit, which had been responsible for assembling the Avias at Aqir, from that vulnerable aerodrome to Ma'abarot, a kibbutz with an airstrip between Netanya and Hadera, about 50km north of Tel Aviv.

Like the Egyptians, the Israelis tried to make operational aircraft from redundant ex-RAF machines, scrap and even parts from crashed aircraft – reportedly including downed Egyptian Spitfires. Whether this really included elements of Barakat's Spitfire 9 remains a matter of controversy. Some Israeli sources have claimed that this was the case. Others note that Barakat's machine had its undercarriage and one wing torn off, and ended up partly in the salty Mediterranean Sea, which would have done its instrumentation and airframe no good. The claim that it was essentially the same machine as the IAF's first operational Spitfire is now widely dismissed as a myth. The Rolls Royce Merlin engine may, however, have been repaired and used to power a different Spitfire that had been rebuilt or reassembled. A talented Israeli aircraft engineer named Freddie Ish-Shalom was credited with this achievement. Zachi Shaul, a mechanic who had previously worked on the Rolls Royce Merlin 24/2 engines of BOAC Lancastrian freighters at Lydda, may have supervised the overhauling of the engine.

Even more important than acquiring aircraft was the IDF/AF's hiring of *Mahal* pilots from Australia, South Africa, Canada, America, Britain and elsewhere. Whether these men are regarded as altruistic foreign volunteers motivated by support for the creation of an independent Jewish state, or simply as mercenaries, is obviously a matter of opinion. The significance of their role during the Palestine War or Israeli War of Independence – again according to where one's sympathies lie – cannot be denied. Indeed, it is hard to understate.

ACTIVITIES DURING THE FIRST TRUCE: THE UN

After failing to produce a settlement of the Palestine problem, the UN Mediator had urged a truce of indefinite length to provide time for further mediation efforts. On 7 July, the Security Council passed another British draft resolution which appealed 'to the interested parties to accept, in principle, the prolongation of the truce for such period as may be decided upon in consultation with the Mediator'. The Israeli government, which needed more time to complete the build-up of its military forces, was willing to accept a limited continuation of the truce. On the other side, the Arab governments found themselves trapped by their own propaganda. The greater part of the Arab population, misled by official claims, believed that the Arab armies were about to achieve a real victory and that Israel was on the verge of collapse. Consequently, street demonstrations and many newspapers clamoured for a resumption of the fighting. Fearing either to reveal the true military situation or to resist such calls, the Arab League Political Committee rejected a continuation of the truce.

Despite this collective decision, not all Arab leaders were willing to renew the fighting; at least not without protest. King Abdullah, for example, welcomed Count Bernadotte's proposals and expressed his willingness to renew the truce, whereas the Transjordanian Prime Minister, Tawfiq Abu al-Huda Pasha, voted to renew the fighting. He later explained his decision as follows: 'All the others wanted to renew the fighting. If I had voted alone against it, we should only have been denounced as traitors, and the truce would still not have been renewed. Jordan (or Transjordan as it currently was) cannot refuse to fight if the other Arabs insist on fighting'.

King Abdullah suspected that the decision of the Arab League Political Committee was taken in order to undermine his own plan to merge what is now the West Bank of Palestine with his existing kingdom east of the river Jordan. This plan was clearly unacceptable to other members of the Arab League. King Abdullah therefore told Bernadotte of his extreme unease at the prospect of war breaking out again, and urged the Mediator to use the full power of the UN to prevent it. The Arab opposition to a renewed Truce therefore increased disagreement between the Arab states and, once the Truce had expired, political suspicions, especially those between Transjordan and Egypt, resurfaced with greater intensity. Furthermore, the fact that King Abdullah had not abandoned his desire to annex part of Palestine, while the Egyptian and other Arab leaders continued to work for the establishment of a separate Arab government for all of Palestine, deepened confusion over Arab political goals.

Meanwhile, Count Bernadotte continued to have exploratory discussions with both sides concerning the future of Palestine. On

28 June, he submitted his first suggestions to the Arabs and the Israelis. Omitting all reference to the UN Partition Plan, he proposed a political union embracing Transjordan and Palestine, which would have comprised two autonomous units, one Arab and the other Jewish. Bernadotte's proposal further suggested some territorial adjustments, as well as the return of all refugees and some limitations on Jewish immigration. Both sides rejected these proposals. Some significant elements within the Israeli political, power and military structures also began to believe that the partitioned territory, as envisaged by the original UN Plan, would not provide the Jewish state with adequate security or defence in depth.

Back at the strategic junction west of Iraq Suwaidan, the UN was meanwhile trying to organise a system whereby each side could cross the area at specific times. It never really worked.

FIRST UN TRUCE: EVENTS DAY BY DAY

Saturday 12 June was the first full day of the Truce, but already concern about the Israelis smuggling war materials across the ceasefire lines prompted the Egyptian military command to request a REAF reconnaissance of the area. This sortie by three Spitfires was not recorded in the al-Arish Operational Diary because they took off from Almaza outside Cairo. They were thus operating at the limit of their range, so the pilots were allowed to land at the RAF aerodrome at Fayid next to the Suez Canal Zone as they returned, low on fuel, after checking Israeli observance of the Cease Fire. It was a strange foretaste of what would happen during the final days of the war. This may also have been the occasion when the Israelis wanted to scramble Avias against REAF Spitfires reported south of Jaffa, but the off-duty pilots were away and by the time they got back to their airfield the scramble had been called off. Whether such long-range reconnaissance sorties were repeated during the First Truce is unknown.

Also on 12 June, the Egyptian Army noted some seemingly random shooting by the enemy. Elsewhere the tedium and unpleasantness of life on the front line during the truce was clear in the diary of Egyptian Lieutenant Salah Badr. He wrote that some of the dead from previous fighting had not been buried, and that there was sickness in the Egyptian ranks, including himself. He was, however, cheered by the arrival of two letters, one of which was from his mother. At this time, Lieutenant Badr seems to have been with the main unit of the Army near Isdud, not in Iraq Suwaidan, where he would later distinguish himself.

In Syria, SyAAF pilot Faisal Nassif's diary still focused upon the fortunately lowering tension within his family caused by his marriage to a Christian girl: 'Alice made a good impression on them and only my father still objects to my marriage'. There would then be a blank in Nassif's diary until 25 June. This was because it was a period of intensive training before being sent to an active squadron.

Through the subsequent days, from 13 to 24 June, Lieutenant Salah Badr's diary continued to record tedium, sickness, flies, scorpions and other troublesome insects. He noted his own increasingly short temper, and was coming to rely upon 'Tuscany' cigarettes. There was occasional excitement and on 16 June he was put in charge of some infantry training.

For most of the time the life of REAF prisoners-of-war in Israeli hands was similarly tedious. However, rather than being repatriated after the First UN Truce came into effect, on 14 June Abd al-Rahman Inan was given an hour to get ready to move to another POW camp. Here, he was told, new clothing, shaving equipment and other 'comforts' would be provided. He was driven no more than 10km north in a civilian vehicle, which then turned left along a dirt track, then up a hill before stopping in front of a wooden gate.

Inan had arrived at what would be his new home, in an area cleared by burning and next to a foul-smelling abandoned village. Unlike the previous camp, this one had watchtowers and floodlights. A more substantial building, which was said to have once been the village mayor's house, stood at the centre and had a massive iron door. The camp commandant seems to have been an old Turk, or perhaps a Turkish Cypriot, who had served in the British Army guarding both Italian and German prisoners of war. He claimed to have learned all the tricks used by POWs. He also warned his charges not to try any such things because this war would be short and the guards would watch their prisoners very closely. Furthermore, if Israeli prisoners were mistreated while in Arab hands, Inan and his comrades would suffer in return. Elsewhere, 16 June would be marked by two alleged Israeli war crimes against Palestinians.

On the 17th and 18th Lieutenant Salah Badr continued to bemoan the tedium of life: 'There is nothing unusual. The day is long and boring, the night is short and tedious'. But, on the following day Badr went to the seaside during an off-duty period. It was the first time and he did not go into the water. On the 20th he was back complaining of boredom, routine, diarrhoea and his inability to sleep at night because of the howling of dogs and occasional gunshots. Two days later Lieutenant Badr's major concern was the non-arrival of an expected parcel of 'Tuscany' cigarettes. Maybe that was why he went into Gaza town on 23 June for the first time since arriving at the front. On the 24th he was allowed a telephone call to his family in Egypt. Judging by the content and tone of Salah Badr's diary entries, increasing tension elsewhere on the ceasefire line did not impact his part of the front where his eagerly awaited care-package seems to have finally arrived on 25 June, much to the Lieutenant's delight.

There was undoubtedly increasing concern within the Egyptian High Command who feared that the Israelis would, sooner or later, attempt to break though the tenuous Egyptian line from al-Majdal to Bayt Jibrin. If they did so in the presence of UN personnel, Egyptian troops were instructed not to open fire. If they attempted such a thing without UN personnel being present, the troops were to stop the Israelis by force. The test came on 25 June, when a south-bound Israeli convoy of trucks, accompanied by the UN Observer Team's Chief of Staff, was held up just south of the prominent and tactically significant Hill 113. Apparently, it was unclear whether food could be sent to the Negev since, according to the original truce agreement, such convoys should only go to a 'municipal area', which the Arabs had understood to mean the city of Jerusalem.

Anticipating such an action by the Israelis, the first recorded REAF reconnaissance sorties from al-Arish during this First Truce were flown on Friday 25 June. In fact, the Operational Diary shows that four sorties were flown during this day. Sortie No.1 was a reconnaissance of the area from al-Majdal to Isdud and then eastward along the front-line as far as Bayt Jibrin. There is no mention of bombs being carried by the two aircraft, a Spitfire 9 number 622, flown by al-Utayfi, and a Spitfire 5, flown by al-Aidarus. They took off at 0435 hours and returned at 0600 hours. Visibility had been good and no hostile fire was encountered. According to the Operational Diary comments column: 'Three vehicles were seen near al-Majdal and many moving vehicles on the al-Majdal – Latrun road. Contacted Zamzam (Egyptian Army HQ) by radio, which informed us that there are no (Egyptian) activities from north to south (hence these vehicles were presumably Israeli)'.

Sortie No.2 was more eventful, and would contribute to a near tragedy during Sortie No.3. Sortie No.2's purpose was a further reconnaissance, this time of the area between Hebron, Bayt Jibrin and al-Majdal. It again consisted of two aircraft, both of them Spitfire

9s, flown by Muhammad Abu Zaid, in number 657, and al-Wasimi, in number 642, who took off at 0540 hours. Visibility was still good, and still no hostile fire was encountered. On their return to al-Arish at 0720 hours they reported that:

> Four lorries, two carrying soldiers and two with (presumed) munitions were seen heading from Bir Nafia (unidentified, probably Be'er Tuvia) towards Yasur where an Auster (type, high-wing monoplane) aeroplane was seen near the al-Nagba (Negba) – al-Majdal (road), with United Nations written on it, and also a Dakota which had landed in Aqir colony (presumably meaning Aqir aerodrome). There was no activity whatsoever on the roads leading to Hebron, Bait Jibrin, al-Falujah, al-Magdal.

Unfortunately Sortie No.2 was still in the air when the two aeroplanes of Sortie No.3 took off at 0635 hours, Spitfire 9 number 624, being flown by Yahya, and Spitfire 9 number 646 by Tahir Zaki. Their task was to carry out a third reconnaissance of a similar area, but starting from Bayt Jibrin then flying to al-Majdal and Isdud. Like the previous reconnaissance sorties, they were not carrying bombs, visibility remained good and there was no enemy firing. Both landed at 0800 hours and reported that: 'Two vehicles were seen on the road leading to Bayt Jibrin and one vehicle (on the road) towards al-Majdal. An Auster aeroplane had landed in the area (map coordinates) 12031202 where it was attacked after landing, and both our planes then returned safely to their shelters'.

Map coordinates 12031202 would indicate that the UN Auster, for such it was, had landed on open ground 1,250m north of the Israeli settlement of Negba, and about the same distance west-south-west of the Palestinian village of Ibdis, whose inhabitants would be expelled by Israeli forces just under a fortnight later.

Several versions of the REAF's strafing on a UN aircraft were published later. Perhaps the first appeared in the *Palestine Post* newspaper, which reported that two REAF Spitfires intercepted a white-painted, United Nations Auster (registration VF502, lately of RAF No.1909 Flight), flown by Lieutenant Colonel Maurice Martin of the USAF, on its way to Haifa. The interception began when the Auster was over the vicinity of Kfar Warburg, north-east of Negba. The Spitfires circled the Auster in a menacing manner, causing Maurice Martin to seek a landing place. According to this clearly inaccurate version, just as the Auster touched down, one Spitfire opened fire, a bullet piercing the fuselage from astern and smashing into the dashboard. However, Maurice Martin was unhurt and was able to continue his flight to Haifa once the Spitfires departed. He made a strong protest on arrival and the Egyptians later apologised, stating that the REAF pilot had believed the Auster to have been an enemy machine. However, this account conflates the actions of the two separate pairs of Egyptian Spitfires.

The second version to be published sought to present the Spitfire pilots' actions in a more favourable light, while adding details not found elsewhere. It appeared in Lieutenant Ali Muhammad Mahbub's article, 'Conditions of Flexibility in the Tactical Air Force', published in the *REAF Magazine* in October 1948. He wrote that, following the start of the First Truce period:

> It was also decided to discover the point of (Israeli) penetration and its direction towards the Tactical Air Force, and that some of its aeroplanes would perform an armed reconnaissance over the border of the Egyptian occupied area. The exploratory sorties began at 0430 and a hostile caravan was discovered and its position and direction were identified at 0520 hours, and thus the operation was carried out. However, something new happened. At 0615 the following conversation took place between a pilot (during Sortie 2, therefore either Abu Zaid or al-Wasini) and the commander of the Tactical Air Force. They called each other using codes that do not need to be mentioned:
> – 'Pilot: Plane [Auster] above our lines. Commander: Do not attack it. Pilot [after hesitation]: The aircraft is a type used by the enemy. Commander: Do not attack it. Pilot [after a while]: Plane (going) in direction of headquarters. Commander: Do not attack it'.
>
> The situation can be summarised as follows: Issue: [1] A pilot during an armed reconnaissance saw an enemy-type aircraft over its lines and [heading] towards the command post. [2] The pilot and the Tactical Air Force commander have explicit instructions to strike any hostile aircraft entering the Egyptian occupation zone. The solution: [1] The pilot performs his duty

An Auster communications aeroplane on loan to the United Nations, but still having its British RAF serial number VF502. It was being flown by Lieutenant Colonel Maurice Martin of the USAF, on Friday 25 June, when he was intercepted by two separate REAF patrols, the second of which opened fire after he had landed near the Israeli settlement of Negba. Published in Al-Ahram newspaper on 21 July 1948. (David Nicolle collection)

without reference to the Tactical Air Force commander, or: [2] The Tactical Air Force commander allows the pilot to strike the aircraft. But none of this happened, there is something in the pilot's sense which caused him to refer to his commander regarding this particular plane, and something similar in the commander of the Tactical Air Force telling him to let this plane alone. And this something is the truth of prophecy and foresight. And very few among the millions of people have the truth of the prophecy and the foresight [that I mentioned] and which prompt their owner to display a negative stand towards the imposed duty, especially in an operation like this one in which there is no blame and no repercussions if the duty was carried out in accordance with the previous orders given.

One of the REAF Spitfire 9s flying over King Faruq's convoy when he visited the Palestinian front early in July 1948 – in a scene reminiscent of that during Abu Zaid's interception of the UN-operated Auster. (Author's collection)

What Lieutenant Mahbub failed to mention was the subsequent strafing of this UN Auster by Sortie No.3. Nor did he explain why Sortie No.3, even if it was already in the air, was not warned about the Auster, which Sortie No.2 had already identified as a UN machine. A cutting from an unfortunately unknown Egyptian newspaper, dated 30 October 1948, offered a sympathetic version supposedly based upon Squadron Leader Abu Zaid's own testimony, and which adds further details:

> On this occasion Abu Zaid flew in front of the UN aircraft to check its identity, in case they were Israelis. Col. Der Meulan (De Meulan? – presumably Maurice Mann's passenger), a Frenchman was one of the UN Truce inspectors, said that it was the most frightening experience in his life. He had to open the cockpit window and wave a white cloth at the Egyptian pilot to show that they were from the UN. So, Abu Zaid came closer and then gave the thumbs up sign and flew off.

The rest of this account appears to be almost entirely fictional.

A consciously heroic version appeared in the form of an illustrated strip-cartoon aimed at younger Egyptian readers. Again, there is no mention of the strafing of the UN aeroplane, though it is perhaps worth noting that Tahir Zaki did not fly any further reconnaissance missions during this First period of Truce, though he would be back in action soon after fighting resumed. Later in his remarkable career, Tahir Zaki earned the nick-name of 'Father of Egyptian fighter pilots' because of his enthusiasm, aggressive spirit and refusal to be downhearted despite the many trials and tribulations which the Egyptian Air Force would endure over the next three decades.

REAF Qa'id Asrab Muhammad Abu Zaid's interception of an aircraft used by the United Nations Truce Observation personnel over Palestine; an episode in which tragedy was avoided, as presented in comic-book form in Egypt after the Palestine War had ended. (Author's collection)

Compared with the REAF's previous two sorties, Sortie No.4 was largely uneventful. It was flown by Muhammad Abu Zaid alone, in Spitfire 5 number 616, taking off at 1755 hours on an armed reconnaissance over al-Majdal, Bayt Jibrin and Hebron. He landed at 1900 hours, reporting good visibility, no interference by the enemy, and having seen no enemy activity on the roads between Beersheba, Hebron, Bayt Jibrin and al-Majdal. The only matter of note was the fact that this flight was described as an armed reconnaissance, whereas the earlier ones had simply been reconnaissance missions.

Some Israeli sources refer to REAF bombings of two settlements near Hill 113 near Negba on 25 June, but these are not mentioned in the al-Arish Operational Diary, so if they took place they must have been launched from an aerodrome outside Cairo. A more likely explanation is that Egyptian Army artillery might have opened fire against this target, perhaps while one of the REAF sorties was in the air.

There was also air activity farther north, with an Israeli Dragon Rapide supposedly being lost near the Israeli settlement of Ben Shemen, 4km east of Lydda. However, Arab sources make no claim in this regard. In Syria, new SyAAF pilot Faisal Nassif's diary announced that: 'The signs of pregnancy are already seen on Alice (they had been married on 24 March 1948). I hope she will enjoy good health throughout the pregnancy period'.

Tensions were still simmering on the Egyptian front on 26 June, but the REAF only flew one reconnaissance sortie, perhaps being cautious because of the previous day's events. This Sortie No.1 was another armed reconnaissance, this time by a pair of Spitfire 9s numbers, 622 and 624, flown by Wahdan and Safwat, who took off at 1535 hours and landed at 1655 hours. The visibility was not as good as the previous days, being described as moderate, and they did not record any hostile fire. The area they had to observe included the road from Bayt Jibrin, though Bayt Khalil to Hebron, as well as Anab (4.5km west-south-west of al-Dhahiriya in what is now the south-western tip of the occupied West Bank, close to the road between Beersheba and Hebron) and Jirin (unidentified). They saw very little movement: 'Two vehicles were observed to the west of al-Falujah (heading?) towards Hebron and nothing else was observed in settlements to the east of the coastal road (between) al-Majdal and Gaza'.

On 27 June, Lieutenant Salah Badr finally got the packet of three letters that he had been fretting about for several days, but admitted that they contained nothing important. Overhead, the REAF would fly two sorties. Sortie No.1 was another armed reconnaissance of the strategically important road from Hebron through Bayt Jibrin and (unidentified) Bayt Jizib to al-Majdal, followed by the coast road south to Gaza. It was flown solo by Mustafa al-Utayfy in Spitfire 5 number 616, taking off at 0730 hours. Visibility was good, though with clouds, and no hostile fire was encountered. After landing at 0850 hours, al-Utayfy reported: 'No activity at all on road between al-Majdal and Bayt Jibrin. No activities in any of the (Israeli) settlements east of the Gaza to al-Majdal (road). Smoke seen in Negba colony'.

Towards the end of the day Sortie No.2 was another armed reconnaissance of essentially the same area, which took off at 1710 hours and returned at 1850 hours. This time a pair of Spitfire 9s were flown by al-Utaufy and Yahya al-Aidarus, in good visibility and noting no hostile fire. They reported: 'No activity around the (east-west) road and no activity on the coastal road between al-Majdal and Gaza. Zamzam (the direct radio link from HQ to aeroplanes in the air) contacted our planes and asked for a reconnaissance of the mainland (the area inland from the coastal road) and Burair and al-Jammamah'.

Egyptian field gun in action during late June 1948. (Egyptian Ministry of Defence)

28 and 29 June were largely quiet on the ground and certainly so for the REAF. On the following day, however, senior and middle-ranking front-line Egyptian Army officers in the Second Brigade were summoned to a military conference. Here they would learn about plans that were being drawn up in the event of renewed hostilities. Amongst them was Gamal Abd al-Nasser, the future Egyptian President. Nasser apparently recalled the surprisingly cautious tone of these proceedings, which he thought contrary to the fighting spirit one would have expected. He was also critical of the plans that were under discussion, though of course he was writing with the benefit of hindsight.

The REAF at al-Arish flew just one sortie on the Wednesday the 30th, this being another armed reconnaissance of the roads linking Hebron, Beersheba and Gaza by a single old Spitfire 5 (number 602), flown by Wafa' Effendi (or Wafa' al-Farandi). Taking off at 1015 hours, finding good visibility and facing no enemy fire, he landed again at 1130 hours to report: 'No activity on the Beersheba-Hebron or Beersheba-Gaza roads'. He also suggested that: 'There are no enemy groups because maybe the road (or roads?) had been cut, and there was no activity in the region of Beersheba'. Meanwhile outside Damascus, SyAAF pilot Faisal Nassif's diary noted that: 'Nothing worth noting happened. Batarmas (an unknown individual who might have been a religious figure) visits us, he will mediate between us and Alice's parents'. Perhaps the latter were as unhappy about Faisal and Alice's mixed Muslim-Christian marriage as were Faisal's family.

On 1 July 1948, the REAF ventured to use very vulnerable Westland Lysanders for its reconnaissance sorties. Either these missions were so important that the risk had to be taken, or the Egyptians were confident that there was minimal danger of interception by Israeli Avia S-199 fighters this far south. It may also be significant that the pilot of both Lysanders was named Waffa or Wafa', presumably the highly experienced man who had flown a Spitfire 5 the previous day. The fact that Waffa had apparently been referred to as *Effendi* on 30 June also indicated that he was an officer of status.

He took off with his observer-gunner named Mustafa Yusuf (or Yunis) in Lysander number 503, which had been W503 during the Second World War, at 1020 hours. This Sortie No.1 was to make a reconnaissance of the Israeli settlements of Be'erot Yitzhak (now Alumim) and Gvar'am as well as the neighbouring abandoned Palestinian village of Burayr. With good visibility and no discernible enemy fire, Waffa and Yusuf landed back at al-Arish at 1140 or 1145 hours to report no activity in any of the named locations. An hour and a half later, the same crew of Waffa and Mustafa Yusuf took off again in Lysander number 504, which had been W504 during the Second World War. This time they headed for Be'erot Yitzhak, Negba, Mishmar HaNegev, a nearby but as yet unidentified Israeli settlement whose name also started with Mishmar, and Beersheba. They returned to al-Arish at 1430 hours, noting that they had seen no vehicles or groups of people on the road between Be'erot Yitzhak and Beersheba.

On 1 July, Lieutenant Faisal Nassif of the Syrian Air Force finally found out what squadron he was going to. His diary entry for this day noted that preliminary flying training would be concluded on 4 July. Then: 'I will join the 1st Fighter Escadron at the Istabal with a First Class, in Class (cohort?) A'. The usual meaning of Istabal or Estabal is cattle-shed, but here it referred to a rudimentary airfield of that name in central Lebanon. From here Syrian Harvards had been operating, primarily in support of the Lebanese Army and the ALA (Arab Liberation Army). His colleagues Ad'ma, Marx and Darbal (?) would be going to the Transport Escadron, while Shinnawi (or Tinnawi), al-Adel and Abara would be joining the 2nd Fighter Escadron, which is believed to have been based at al-Mizzah outside Damascus.

On 2 July, the Nassif family entertained some relatives to lunch, perhaps because the prospect of the young pilot being thrown into action within the next few days had soothed family tensions. On the 3rd, Nassif's diary returned to his duties as a husband and soon-to-be father: 'I am thinking of taking Alice to live at Shtura (then a small town on the road between Damascus and Beirut) or at the Istabal (airfield) itself. But she is very tired from the pregnancy'. On the 4th, Nassif's diary proudly announced that: 'The flying training is over and we will start the military training before we join the escadrons. We went home and found Sami and Batras. They brought clothes to Alice'. Clearly, he was not yet ready for active service combat, but on 5 July: 'We commenced the military training with Mato (the Croatian pilot recruited by the Syrian Air Force) we commenced with the --- (unclear) exercise'. His diary entry for the following day, 6 July, strongly suggests that he was still training outside Damascus: 'Everything OK. In the evening we went to the cinema at six o-clock, the film lasted four hours. We got back at eleven o'clock'. That would soon change, although on the 7th Faisal Nassif's was still completing advanced training: 'Alice went into town (Damascus) to see her dress maker and returned to Sa'asa'a (the village south-west of the airfield at al-Mizzah where Faisal and his wife were probably living) at six in the evening. I had an argument with the Yugoslav pilot Mato'. This was despite the fact that Nassif held the Croatian in high esteem as one of the few foreign mercenaries to carry out operational sorties at the front. Dukovac was also known to have criticised the Syrian pilots for their low professional abilities, which led to frequent arguments. The Croatian would, in fact, later leave for Canada, probably during the Second Truce.

On the central front, where Iraqi troops and the Transjordanian Arab Legion faced the Israelis, the latter apparently lost an aircraft to enemy groundfire on 7 July, though the date in not entirely certain. Reportedly, an Auster A-9 of the Negev Flight was flying northwards during the afternoon of 7 July. Near Lydda it was shot down by ground fire, the pilot Amnon Berman and his female passenger being killed. Their remains were recovered by troops from the Arab Legion, who conveyed them to Lydda from where they were collected by the Red Cross and taken to Tel Aviv. That same day the Palestinian village of Tal al-Safi was seized by Israeli forces shortly before the UN Truce was supposed to come to an end. Its inhabitants were expelled on the orders of Shimon Avidan, the commander of the Givati Brigade who had told his soldiers 'to destroy, to kill and to expel refugees encamped in the area, in order to prevent enemy infiltration from the east to this important position'.

Meanwhile, King Faruq of Egypt paid a formal visit to various front-line units on this, the last day before the end of the UN Truce. It was intended to be a morale-boosting tour, and the king was accompanied by Egyptian journalists as well as a news film team. The idea that Israeli soldiers were trying to kill the king when they fired shots in his general direction seems unlikely, as they were fired from long range. On the other hand, given the fact that Israeli intelligence sources appear to have known all about King Faruq's tour, including the timing of his itinerary, these shots are unlikely to have been coincidental. They were more likely to have been intended to 'make a point'.

This tour also included a visit to the REAF Tactical Air Force at al-Arish, during which an unidentified aircraft, believed to be Israeli, reportedly flew over the aerodrome. This was presumably again to make a propaganda point, while no attempt was made by the Egyptians to intercept it.

Pilots of one of the SyAAF's Fighter Escadrons, with one of their mounts, either during a graduation or establishment ceremony, in 1948. (Albert Grandolini Collection)

REAF pilots and some ground-crew around an Egyptian Spitfire 9 as part of their role as 'extras' in the film 'A Girl from Palestine'. The aeroplane itself clearly shows signs of hard wear in Egypt's dusty and unforgiving climate. (Author's collection)

Most sources agree that the morale of REAF aircrew remained high, though their most senior officers were not so confident. Egyptian news-film coverage of King Faruq's visit to al-Arish aerodrome included several interesting features. Amongst these were a distant view of the Hawker Fury prototype and a closer view of its undercarriage; neither of which show much of the machine's marking, although the original red fuselage stripe had clearly been removed. There was also a scene of Squadron Leader Muhammad Abu Zaid, with a Spitfire 9 bearing an impressive list of targets attacked and damage inflicted painted beneath the cockpit. Considerable effort has been made to interpret this list of targets and claims. The different sized stencils of armoured vehicles might indicate tanks and armoured cars. The Arabic word *Matar* meant an aerodrome of any size, while the word *Mustamid* meaning 'camp' probably indicated a fortified kibbutz or other such Israeli settlement. Nevertheless, the following list is by no means confirmed:

6 small aircraft, which look more like Gloster Gladiators than anything else.
Matar (airport) *Aqir* – 2 bombs and 2 large aircraft, which look like Dakotas.
Matar (airport) *Tel Aviv* (Sde Dov) – 11 bombs and 1 large aircraft, which looks like a Dakota.
Matar (airport) *Bayt H-k-ma* (unidentified but probably Nir Am or the ex-British airfield just south of Tal Abu Hurairah on the road between Gaza and Beersheba) – 3 bombs.
Mustamid (kibbutz) *Shaydun/Shaydum* (unidentified, but probably the Negev Brigade Base near Nir Am*)* – 5 bombs and 3 armoured cars.
Mustamid (kibbutz) *Kfar B-r-n-r* (probably Bitzaron) – 5 bombs and 3 armoured cars.
Mustamid (kibbutz) *Negba* – 4 bombs.
Mustamid (kibbutz) *Dayr Sunayd* (Yad Mordechai) – 5 bombs.
Mustamid (kibbutz) *Dangum* (Danjur?) – 2 bombs.

A middle-distance view of one of the REAF Lysanders based at al-Arish merely confirmed that they were there; while the REAF's own emergency fire service, not to be outdone, featured resplendent in their somewhat incongruous 19th century style fireman's helmets.

According to official Egyptian records, the operational strength of the REAF on 7 July 1948, at the end of the First UN Truce, now stood at 50 aircraft, which is four more than it had been at the start of the Truce. A more detailed breakdown and comparison with 11 June shows the following:

Fury: x 1 on 11 June; x 1 on 7 July.
Spitfire 9 fighter/fighter-bomber: x 18 on 11 June; x 16 on 7 July.
Spitfire 5 fighter/fighter-bomber: x 9 on 11 June; x 14 on 7 July.
Lysander: x 2 on 11 June; x 2 on 7 July.
Hurricane (Met-Flight): x 3 on 11 June; x 5 on 7 July.
Dakota transport-bomber: x 4 on 11 June; x 7 on 7 July.
Dove transport: x 4 on 11 June; x 5 on 7 July.
Anson transport-bomber: x 0 on 11 June; x 0 on 7 July.
Commando transport: x 2 on 11 June; x 3 on 7 July.

The main changes, apart from a slight increase in the overall total, was a drop in the number of Spitfire 9s, probably because they needed a major overhaul after very intensive use, and a substantial increase in Spitfire 5s. These had largely been held in reserve, but were now serving at the front on an equal footing with Spitfire 9s. It appears that perhaps all currently serviceable Hurricanes were used by the Meteorological Flight work whose role had been given greater priority. There was also a reduction in the number of operationally available Dakota bombers, probably because these previous redundant or 'scrap' machines had been used intensively during the First Phase of Fighting. The REAF had acquired a newly delivered Dove, probably using it for ordinary communications and transport work, while one more 'rebuilt from scrap' Commando was also available.

This list of operational aircraft did not, of course, including training machines, and perhaps not the Royal Flight, which still officially formed part of No.3 Communications Squadron. Quite where the additional Westland Lysanders that still formed part of the REAF's inventory fitted in is unknown.

3

SECOND PHASE (8 TO 18 JULY 1948)

The Muslim Holy Month of Fasting, Ramadan, began on Thursday 8 July 1948 (the Islamic or Hijri year 1367 AH), and would last until 5 August. Whether this had any bearing on the timing of the start of the Second Phase of Fighting – the so-called Ten Days War, because it started during the night of 8-9 July – is unclear, but seems highly likely. The UN Mediator in Palestine, Count Bernadotte, had worked hard throughout the First Truce, trying to get all sides to agree to an extension. He had shuttled between Cairo, Damascus, Amman, Beirut, Tel Aviv and Haifa to discuss the situation with Arab and Israeli leaders, with little to show for his efforts. Time was, in fact, running out for the First UN Truce.

FORCES AND STRATEGIES INVOLVED

The Israelis knew full well that Bernadotte's plan would give them less than the original UN Partition resolution had offered. So, the Israeli government not only rejected the plan, but actively worked towards its collapse. In order to do this, and to achieve their most important objectives as rapidly as possibly, the Israelis formulated a new strategy. First, they had to secure New Jerusalem and their control over the access routes as firmly as possible. Once Jerusalem was protected, they then had to open and secure supply routes to the Negev settlements. Elsewhere, they planned to eliminate the Syrian bridgehead and drive out Quwaqji's ALA from upper Galilee, while nibbling away at all Palestinian areas still under Arab control and to prevent the Arabs from expanding any of their current holdings.

Once the UN Truce ended, the Israelis were, for the first time, in a position to take the offensive on several fronts simultaneously, although the main Israeli effort took place on the strategically vital central front. On the other side the Arabs, who rejected a continuation of the truce, adopted a defensive strategy, which not only handed the initiative to the Israelis and encouraged them to attack wherever they liked, but also made it impossible for Arab armies to gain more territory or even to keep all of what they had secured during the first round of fighting.

EGYPTIAN FORCES

Immediately after the Second Phase of Fighting erupted, the commander of the Egyptian Expeditionary Force in Palestine, Major General Ahmad Ali al-Muwawi, was urged to go on the offensive to relieve pressure on his Arab allies. This he was prepared to do, but not before warning the Egyptian government that his shortage of reliable ammunition would not only make it virtually impossible to maintain such an offensive, but would put his troops in danger if they were obliged to retreat. In fact, al-Muwawi demanded that someone qualified should inspect the Expeditionary Force to verify that his fears were justified. In practice, the Egyptians, having been profligate in their use of artillery during the First Phase, now found themselves obliged to rely on inferior artillery shells. Al-Muwawi's claim was also verified from the Israeli side, where it appeared that the Egyptians were economising on their use of artillery and had virtually ceased counter-battery fire. Many of the shells that they fired were duds, and some were found to be up to 30 years old.

Another result of such shortages was that Egyptian representatives were increasingly active in the 'open – or more correctly the illegal – arms market. They were not very successful and one of the most notable results was the series of 'arms scandals' that rocked Egypt after the Palestine War ended. Abd al-Latif Baghdadi of the REAF, and subsequently a senior member of President Nasser's government, wrote that most such black-market equipment and munitions dated from the Second World War, was limited in scope and inferior in condition. Baghdadi also pointed out that this was in stark contrast to the high quality equipment that the Israelis were obtaining from Czechoslovakia and elsewhere.

Nevertheless, the Egyptian Expeditionary Force had been increased in size to about a Divisional Group, while al-Muwawi moved his command post from Gaza to al-Majdal. Thus, by the end of the truce, Egyptian troops were deployed in a static defensive array in seven defined sectors. About one brigade was deployed along three coastal sectors from al-Arish to Gaza. Another brigade was deployed along two lateral sectors from al-Majdal to Iraq al-Manshiyah, while a third brigade was deployed in the Isdud sector. Meanwhile, along the internal axis, the Light Force still largely consisted of Arab volunteers supported by regular Egyptian Army infantry. It numbered around a battalion and occupied key positions along the desert road from al-Auja on the Egyptian frontier to Bethlehem.

Before the Ten Days fighting broke out, some British and American observers declared that in any future fighting the REAF would not even be able to defend al-Arish let alone conduct offensive operations over Palestine. Their warnings were premature, though some new pilots had been sent to al-Arish with minimal training. Amongst them was Kamal Zaki who recalled that, having done his advanced training at Hilwan, he arrived at al-Arish about a month and a half after the war started, which would have been during the First Truce. Not all the REAF's more experienced pilots had been sent to al-Arish before the war started. For example, Squadron Leader Mustafa Kamal Nasr Abd al-Wahab seems to have been first mentioned in the Operational Diary on 12 July 1948. Prior to that he was one of the youngest REAF pilots to be given important responsibilities, including command of the Meteorological Flight.

Qaid Agrab Mustafa Kamal Nasr (killed in the Palestine War, late on). (Dr David Nicolle collection)

After that, Mustafa Kamal Nasr served as a flying instructor in charge of 'combat formation' training. He was now sent to al-Arish, probably at the head of a cohort of newly trained pilots.

Kamal Zaki, when interviewed by the author in 1999, recalled that he and his young colleagues soon found themselves very busy. He himself first appeared in the al-Arish Operational Diary on 12 July, the fourth day of the Second Phase of Fighting, and recalled that: 'The main opposition were the Israeli Messerschmidts (Avia S-199s). I flew Spitfire 5s and Spitfire 9s. REAF Spitfire missions from al-Arish were designated as ground-attack, or air combat interceptions, or convoy attack missions. We flew in pairs, one aircraft to attack and one to defend him'. Kamal was fully aware of their limitations, telling the author that: 'In the early days there were very few of us, and we used to be called The Fifty Fighting Pilots. We were not very well trained, nor were we well equipped'. Kamal Zaki also added, almost as an aside: 'Personally I think that (Muhammad) Abu Zaid was overconfident'. Other men mentioned during this period were Yahya al-Shanawi, who served as a Senior Staff officer al-Arish, and Abd al-Bakr Mawsi (Mursi), a multi-engine pilot based at Almaza in 1948, who later rose to the rank of Lieutenant Colonel in the United Arab Republic Air Force (UARAF; official designation of the Egyptian Air Force from 1958 until 1972).

Back in Almaza, the training of new cadet pilots proceeded as rapidly as possible in the existing difficult conditions. At Hilwan others started their advanced training on new Italian fighters. Advice would be sought from experienced front-line fighter pilots whenever the chance arose. On one such occasion Tahir Zaki was on leave in Cairo when he received a telephone call from Group Captain al-

Officers' Mess at the Almaza Air Base. (David Nicolle collection)

Miqaati, who asked him to speak to a group of new pilots before they left by train from Cairo to al-Arish. As Tahir Zaki recalled: 'I told them to try and attack the enemy head-on because this always made the Israelis turn away. As I remember, the train waited until I had finished talking to these pilots. I continued to fly Spitfires while these new pilots were given new Fiats and Macchis'.

Shalabi al-Hinnawi's recollections of this period contained an otherwise uncorroborated reference to a two-seat Spitfire conversion trainer:

> I graduated in 1948 and went as a Lieutenant to the Flying School, it wasn't yet an Academy, at Almaza. There I was always top of my class. It was the same when I was training in England and Russia. In fact, I started to think in English. The instructors at Almaza had themselves been trained at Abu Suwayr (by the British). We flew Magisters and Harvards. The Harvards had come from the USA. Then we went back to flying Magisters again. Fighter training was done in the same School on a two-seater Spitfire. The REAF originally had two of these (British records indicate that only one, serial number 684, was sold to Egypt in 1950). Most of the new pilots went straight to single-seater Spitfires, instead of having experience on the two-seater. The REAF also started a twenty-hour training and a conversion programme. I myself never locked the hood of my Spitfire. It was so old that it was difficult to see through and it also tended to jam shut. In one aircraft the two lamps which were supposed to show that the wheels were down and locked had been joined up the wrong way around, so up was down and down was up. I ended up making a belly landing when the lights showed that the wheels were supposed to be down.

In a subsequent on-line account of his flying career, 'Eagles of the Air in Palestine', al-Hinnawi wrote that:

> Then we had our final exam and we graduated in July 1948 in a very simple ceremony, and I was the first on the cohort and we were 14 out of the original 26 (cadets). After graduation we returned to the Higher School for Aviation with the purpose of flying on Spitfire fighter airplanes after receiving – on the ground – information on the aeroplanes from more senior pilots. It was sad and shameful that, when studying the aircraft itself, we found that half the airplane's instruments did not work and the plane's hood, if closed, obstructed vision but was unsafe if open. And the aircraft itself was combined from scrap, the body had a number, the right wing had a number, and the left wing another number, i.e. it was combined from 3 airplanes. As a result, several times the propeller separated from the aeroplane during flight, and also the control column, which happened with Air Comm. (a rank later in his career) al-Talti during his flight. Despite this, we had our first flight on these machines, but our flying time did not exceed 5 hours, just to familiarise us with the aeroplane, and all were like suicide flights, the pilot did not know if he would be successful in landing or not!!!!

IRAQI FORCES

Far less is known about the other Arab forces committed during this period, though the Iraqi Army contingent had increased to two infantry brigades deployed in a triangle of territory between Jinin, Tulkarm and Nablus. In fact, the Iraqi Army did not receive the military equipment its government had ordered from Britain

King Faruq of Egypt leaving al-Arish at the end of his official inspection of the airbase, on 7 July 1948. Among the saluting senior officers is Muhammad Abd al-Muna'im al-Miqaati [second from right], whose Westland Lysander would be attacked by an Israeli Avia S-199 two days later, resulting in the death of the Israeli pilot, Bob Vickman. (Author's collection)

until well after the Palestine War ended. This is said to have been one of the main reasons why Iraqi forces remained largely inactive. The embargo also hit the RIrAF, though perhaps not quite so hard. The arrival of the first few Hawker Fury fighters was welcome and potentially significant, 10 of the promised 34 machines being delivered during 1948. On the other hand, in July the British government refused to send further spares for grounded Iraqi Furies. Furthermore, the RIrAF still lacked bombers or crews capable of carrying out night attacks.

Thus, at the start of the Second Phase of Fighting, the RIrAF still had some Ansons based in Transjordan, though the Gladiators had either been withdrawn or would no longer be used. In the absence of official or reliable information to the contrary, it currently seems most likely that the first Iraqi Furies to be sent to Mafraq in northern Transjordan arrived on 16 July, towards the end of the so-called Ten Days War, at a time of considerable urgency.

According to General Arif Abd al-Razzaq, who was a Lieutenant in No.7 Squadron RIrAF at this time, six Iraqi Furies were sent to the front. However, the Syrians' need for air cover was clearly greater than that of the Transjordanians, so these Iraqi Hawker Furies were sent to Damascus within a few days – if not hours – of their arrival. Arif Abd al-Razzaq recalled that: 'They arrived in Damascus around noon, but one crashed on landing. The pilot survived and was safe. Some ground personnel were already there, with spares and 6,000 rounds of 20mm ammunition, almost all the RIrAF had for its Furies at the time'. Quite how many were sent to Damascus also remains unclear, as Egyptian official sources maintain that two Hawker Furies were on the REAF's strength on 18 July, at the end of the Second Phase of Fighting. If this is correct, then it would seem that one of the Iraqi Furies had already been sent to Egypt for evaluation.

In an arrangement between Baghdad and London from December 1946, the RIrAF was to acquire 30 Hawker Sea Fury FB.Mk 11s (designated Fury F.Mk 1 but colloquially known as 'Baghdad Fury'), and five (later reduced to two) Fury Trainer T.Mk 52 conversion trainers. Shown here is one of – eventually – two Fury conversion trainers to reach Iraq, the serial number 263. (Albert Grandolini Collection)

The first eight single-seat Baghdad Furies, and one two-seater had reached Iraq in November 1947. Visible here are the first three single- and one two-seaters prepared for Iraq, at the start of their journey to the Middle East. Their serial numbers were (from front towards the rear): 233, 232, 231, and (two-seater) 238. (Albert Grandolini Collection)

The conversion of crews from Nos.1 and 7 Squadrons, RIrAF was still going on at the time of the First Phase of the Palestine War. Visible here during a training sortie over northern Iraq are the examples with serial numbers (from front towards the rear) 249, 247, and 248. (Albert Grandolini Collection)

In the general atmosphere of 'quiet instability' and while still busy demobilising and repatriating dozens of thousands of military personnel left over from the Second World War, the RAF units assigned to the Air Headquarters Iraq and Persia (colloquially 'AHQ Iraq') were partially re-equipped with Hawker Tempest fighter-bombers. This example was operated by No.249 Squadron from Habbaniya in Iraq. (Albert Grandolini Collection)

SYRIAN FORCES

Meanwhile, the Syrian bridgehead at Mishmar HaYarden was now defended by the 1st Infantry Brigade supported by the Syrian 2nd Brigade, positioned on the heights overlooking the east bank of the Jordan River. The Syrian government, which was relatively prosperous at this time, also turned to Czechoslovakia with a request for arms, but unsuccessfully at first: Prague was too busy supporting Israel. According to Faisal Nassif's diary, as of 14 July 1948 the Croatian top fighter ace of the Second World War, Mato Dukovac, was still in Syria, flying for the SyAAF. However, he is understood to have left the country later in the year, migrating to Toronto in Canada where he lived until his death, on 6 June 1990. Four other Croatians continued serving with the Syrian Air Force before, in 1950, being contracted by Israel, where they trained pilots of the IDF/AF for about a year, before travelling to Italy. The American journalist John C. Ross, in an article written for the American magazine *Flying* (October 1948), claimed that two French pilots had recently been enlisted by an unnamed Arab Air Force. There is no evidence for French personnel in the SyAAF so, if true, Lebanon seems a more likely candidate.

Other sources have suggested that by this time the Syrians only had 12 serviceable Harvards, but the claim that they were only used when Syrian skies were threatened is clearly false. More probable is the suggestion that the SyAAF did acquire an unknown number of Avro Ansons in early July 1948, these being the last addition to the SyAAF until after the Palestine War. Where they came from is currently unknown. A loan agreement with Iraq seems most likely, though the REAF has also been suggested as a possible source. According to the Israeli aviation historian Noam Hartoch, the two Ansons used by the SyAAF as bombers from mid-July onwards were fitted with external racks. They also reportedly still had upper gun turrets with a 0.303 inch machine-gun. A second such gun, which is also mentioned, would have been in a fixed position in the front fuselage. Whether these were the same two Ansons that had been obtained as navigation trainers is unknown. The only serviceable Syrian Airlines C-47 Dakota had already been commandeered by

Four youthful pilots of the Syrian Arab Air Force, as seen at Aleppo/Nayrab Airport, together with one of AT/T-6 Harvards. (Albert Grandolini Collection)

A scene from the graduation ceremony for a new class of SyAAF T-6-pilots, at Aleppo/Nayrab airfield, at an unknown date in 1948. Relatively wealthy at the time, Syria invested heavily into building-up an air force, and training additional flying- and ground crews, but was badly hampered by British arms embargos, or refusals of other countries to sell it the necessary equipment. (Albert Grandolini Collection)

the SyAAF, being used to ferry officials to and from Egypt, and to carry munitions, usually bombs and small arms ammunition, from Egypt to Syria. It is also said to have been captained by a British pilot, and that the British government did not complain about this because the flights were between Arab states and therefore did not violate the arms embargo.

LEBANON, TRANSJORDAN AND PALESTINE

At the end of the First Truce, the Lebanese army was still deployed in defensive positions along the Palestinian-Lebanese borders, and between the coast and al-Malkiya. It could also rely on some degree of air support from the Syrians.

There was similarly little change in Transjordanian deployments, and the country had some degree of air cover from the British RAF, though much less than some pro-Zionist elements in the world-wide media claimed. Furthermore, it only applied to Transjordan itself: namely east of the river Jordan. In this context, the words of the American journalist John C. Ross might be worth quoting:

> The Transjordan Air Force (meaning any aircraft used to support the Arab Legion, as there was as yet no Transjordan Air Force) is completely run by the RAF and the entire cost of its operation is paid by the British crown … After one raid on Tel Aviv the Jews retaliated by a counter-raid against Amman, capital of Transjordan. The British immediately notified the Jewish authorities that Amman was a British military stronghold and that if there were any further raids on Amman by the Jews, the RAF would shoot down any Israeli planes found in the skies.

Meanwhile, the unfortunate indigenous Palestinian militias, including those of the Arab Liberation Army, had become a virtual irrelevance in military terms. Nevertheless, an ALA force of nominally around 2,200 men still held most of central Galilee from Nazareth to the Lebanese border.

ISRAELI FORCES

Things had, of course, changed much more dramatically on the Israeli side. In addition to a significant number of tanks, other armoured vehicles and numerous if somewhat varied artillery, Israel now had a substantial, moderately well equipped and well-trained air force. As a consequence, the Israeli High Command was confident of its ability to defeat its foes in the north; namely the Syrians, Lebanese and remaining elements of the ALA. Against these northern Arab forces, the Israelis had deployed five brigades along the truce lines in north-central Palestine. These were the Alexandroni, Golani, Carmeli and Oded Brigades, plus the 7th Brigade, which had been transferred from the Latrun sector to western Galilee. In addition, there were still local defence garrisons.

Unfortunately for the IAF, after Czechoslovakia's supply of Avia S-199 fighters to Israel became widely known the Czech government were reluctant to supply the reconditioned Spitfires which Israel had also ordered, at least until secrecy could be guaranteed. Meanwhile a significant number of the IAF's Avias had been destroyed or damaged, though mostly not as a result of enemy action.

ARAB STRATEGY

President Nasser's subsequently published account of his own experiences during the Palestine War can be confusing where precise dates are concerned. He was nevertheless highly critical of the way in which Egyptian Army units were scattered along fronts that were too long for their limited numbers to hold, or even sometimes to patrol. He was also painfully aware of the Egyptians' lack of sufficient reserves, although nothing could be done about that. The results were piecemeal attacks, which caused unnecessary and sometimes almost pointless casualties. Nasser would, of course, be writing in justification of his own 'revolutionary' government's criticism of its 'royalist' predecessors. The Egyptian Army continued to suffer from inefficient logistics and an officer corps that had not really been trained to cope with rapidly changing battlefield conditions, though there were of course exceptions, not least in the Air Force.

The Egyptian High Command's plans for a limited offensive during the Second Phase of Fighting aimed at widening the coastal and lateral sectors that the Egyptian Army already controlled, and to tighten their grip around Israeli settlements in the Negev. This would entail capturing several strongholds and outposts north and south of the al-Majdal to Bayt Jibrin lateral road. The result was a number of localized attacks that achieved limited success. On the other side, comparable local Israeli offensives resulted in some successes and some failures, but were in general more satisfactory than Arab offensives. Furthermore, from the Second Truce onwards the Palestine War became primarily an Egyptian-Israeli one.

The Egyptian plan gave the major tasks to Brigades currently based in, around and defending al-Majdal and Isdud. The 2nd Brigade Group, in the Ashdod-Nizzanim sector, was commanded by Brigadier Mahmud Fahmy Ni'mat Allah and was tasked with extending the ground it held eastward to a line drawn between Isdud, Bayt Daras, Sawafir al-Sharqiyah (Eastern Sawafir) and Julis. The 4th Brigade, now commanded by Brigadier Muhammad Naguib, was tasked with widening the lateral sector southward by seizing the line east of Iraq Suwaydan, Kawkaba, al-Hulaiqat and Burair. In fact, Egypt's four brigades were now all fully committed and any offensive force had to consist of one or two battalions drawn from the al-Majdal or Isdud Brigade areas.

On the other side, the Israeli plan named *An-Far* (for "anti-Farouq") called for a coordinated strike by the Givati and Negev Brigades to clear the road to the Negev, to be launched during the night of 8-9 July. It aimed to cut the Egyptian-held lateral east-west supply route and also to force the Egyptians out of Isdud. Colonel Shimon Avidan, commander of the Givati Brigade, was appointed as the overall operation commander. During the first night, the Givati Brigade was to conduct mopping-up operations to clear any Arab villages and positions remaining within its existing territory, thus securing the flanks and rear. That same night, the Negev Brigade was to raid Arab villages in the vicinity of Kfar Darom, within what is now the Gaza Strip, to relieve the defenders of that Israeli stronghold. During the second night, the Negev Brigade was supposed to capture the Iraq Suwaydan Police Fortress, while the Givati Brigade was to operate against a number of nearby Arab villages to keep the Egyptians and local Arab militias busy some distance north.

By the end of these operations, it became clear that the Egyptian Army had reached the limit of what its supplies and capabilities allowed. However, although some Egyptian forces achieved limited local success, the overall Israeli plan failed. Even with that being so, the Israelis generally had greater reason for satisfaction than did the Arabs.

When fighting resumed after the First Truce, combat actions by the Syrian and Iraqi Air Forces were limited. Due to their shortages, they had to focus upon supporting ground troops. Hence, except for one possible Iraqi sortie against Ramat David aerodrome, both the RIrqAF and the SyAAF refrained from action against enemy airfields because they were not in a position to face the Israeli Avia S-199s that might rise to challenge them.

In the central sector, where the Israelis faced the Arab Legion to their north and the Egyptian Light Force to the south, the Israeli Army still controlled a corridor of territory linking the coastal Plain and Jerusalem. But, they did not control all of the existing main road between these two areas. The Israeli High Command therefore ordered the construction of a new track capable of being used by motorised vehicles south of the existing one. It started near the village of Dayr Muhaysin, held by Zionist-, now Israeli, forces since December 1947 when its Arab inhabitants were expelled. The new road then climbed some rugged territory, south of the similarly 'cleansed' Palestinian villages of Bayt Jiz, Bayt Susin and Bayt Mahsir, before rejoining the old Jerusalem road close to the 'cleansed' Arab village of Saris. This route came to be popularly known as the Burma Road in memory of the construction of a road between British-ruled Burma (now Myanmar), the north-eastern tip of British-ruled India and allied China. Its successful completion would enable supplies to be sent unchecked by Arab forces or by UN Truce Observers.

Farther south, the IDF General Staff had deployed two reinforced brigades (the Givati and Negev) on the southern front during the truce, in addition to the existing settlement garrisons in the Negev. The Givati Brigade was responsible for a wide arc of settlements and outposts north of the road linking al-Majdal and Bayt Jibrin, from Gal On in the southeast to Yibna in the northwest, close to the extensive sands that marked the Palestinian coast. These forces consisted of two full battalions plus settlers now turned full-time soldiers, with another two battalions held in reserve. To the south, two of the Negev Brigade's battalions were responsible for an even larger area in which 26 Israeli settlements were widely dispersed north and south of the Gaza-Beersheba road. A third battalion was kept free to go wherever it was most needed.

The Douglas DC-3 of BOAC in which King Abdullah of Jordan arrived for a visit to Egypt, on 22 June 1948. (David Nicolle collection)

4

CHRONOLOGY OF THE FIRST THREE DAYS OF THE SECOND PHASE (8 TO 11 JULY 1948)

The Second Phase of fighting started in the north with an Israeli offensive against Syrian troops holding Mishmar HaYarden, and in the central sector facing the Jerusalem Corridor targeting Transjordanian forces in the Ramlah and Lydda areas. In overall command of Israeli forces in the north was Brigadier General Moshe Carmel, who was charged with dislodging the ALA from the strategic positions it still held in central Galilee. To achieve these objectives, two offensives were planned, Operation Brosh against the Syrian bridgehead, which was less than 2km from the vital north-south road across eastern Galilee, and Operation Dekel against the ALA. The scene was thus set for some vicious though relatively brief fighting on most fronts.

According Andrew James McGregor in his book, *A Military History of Modern Egypt*, on 6 July, the combatant Arab nations had decided to resume the war. On the Egyptian front the Israelis began their preparations for an offensive by seizing most of the Arab villages and key positions that were needed for securing this operation. However, the Egyptians could see what was going on and promptly notified the UN Mediator. As the Egyptian Command did not receive any reply from the Mediator until 7 July, General al-Muwawi decided to pre-empt the anticipated Israeli attack by launching the Egyptians' own offensive on the 8th, seizing the initiative and gaining some element of surprise. This day was also the first day of the Muslims' holy fasting month of Ramadan.

8 JULY 1948
GROUND OPERATIONS

At dawn on 8 July (0440 hours), after a brief artillery bombardment, a Sudanese infantry company commenced the Egyptian attack on Bayt Daras, a short distance southeast of Isdud. According to the Egyptian plan, this Sudanese company was to advance as a spearhead, enabling the main attacking force to break into the village. After fulfilling its mission, it had to fire a green signal so that the rest of the attacking force would advance to secure the village. In case of failure, it had to fire a red signal, as a request for artillery support. At 1000 hours, the Sudanese company successfully completed its mission but instead of firing a green signal it fired the red one. Colonel Abd al-Hamid al-Habruk, the commander of the attacking force, responded with artillery fire on the Israeli position which was now occupied by the Sudanese company. Now suffering losses from friendly fire, the Sudanese were compelled to withdraw and the seized positions were quickly retaken and fortified by the Israelis.

An ancient field gun of the Egyptian Army, as seen in action in Palestine. (David Nicolle collection)

However, a different account of the struggle for Bayt Daras states that the Sudanese were eventually driven back during a ferocious exchange of grenades. Colonel al-Habruk made another attempt to seize the village but this time the attacking force was unable to break through the strengthened Israeli positions.

Elsewhere, at 0600 hours the Egyptians launched as heavy an artillery barrage as they could, targeting the Israel settlement of Negba. This was followed by an infantry assault supported by armour on both flanks, but the fortifications of Negba had been strongly reinforced during the Truce and the attack was halted. The Israelis then counter-attacked, pushing back the original Egyptian line a little in the centre.

South of the Negev road junction, the Egyptian 2nd Battalion, commanded by Colonel Ahmad Tawfiq, had greater success and, at 0700 hours on the 8 July, it captured Kawkaba from the Negev Brigade. The Israelis withdrew to al-Hulaiqat, but the Egyptians also raided southwards from the lateral road, Israeli troops of the Negev Brigade being taken completely by surprise and chased south by Egyptian armoured vehicles right through al-Hulaiqat village. The Israelis then withdrew to the abandoned Palestinian village of Sumsum, west of Burair. Around the same time a Sudanese volunteer unit, released from the Isdud area, struck northwards from the strategic junction and seized Hill 113. This they hurriedly fortified and manned with infantry, anti-tank guns and support weapons.

The Israeli Command decided that their own planned offensive on the southern front must now be pushed ahead. The Givati Brigade had cleared the Arab villages in its area of their inhabitants, as a result of which around 10,000 Palestinians fled towards Masmiya al-Kabira and farther afield. Seemingly for the first time, Egyptian Lieutenant Salah Badr provided a relatively clear account of military affairs in his diary entry of 8 July, referring to the successful seizure of Kawkaba, the Egyptian Army's near success at Julis and Bayt Daras, followed by a night of artillery fire.

The night of Thursday-Friday, 8-9 July 1948, would be a busy one for many of the men at the various fronts. In response to the Egyptian offensive, the Israelis advanced the timing of Operation An-Far, meaning that units of the Givati and Negev Brigades hastily reorganised and launched their own planned attack earlier than originally envisaged. Operation An-Far would continue until 15 July.

Just after dark, these Israeli troops began to move toward their assigned objectives, Givati units taking advantage of their positions on hills overlooking Ibdis, which had been occupied before the cessation of the First Truce. Bayt Affa, which was defended by a few GMA irregulars, fell after a brief skirmish. However, Ibdis which was also defended by some GMA irregulars was only taken after a long hard fight because the two Israeli attacking platoons, of whom all but 18 became casualties, bumped into a substantial Egyptian force which was said to be preparing to attack Sawafir al-Gharbiyah and Sawafiya al-Sharqiyah.

Other information indicates that this small Egyptian force had attempted to reach the men still holding out in Ibdis by attacking south from the coastal road through Bayt Daras and Julis, and north via Bayt Affa. Egyptian tanks then shelled Ibdis from close range, but one was hit by a 65mm anti-tank gun, so the rest withdrew into Bayt Affa. Salah Badr recorded the Israeli seizure of Ibdis in his diary, and also mentioned that a certain Lieutenant Rifat had been away on leave, probably in Cairo. Whether these two comments were related is unknown. In fact, Israeli mopping up operations in Ibdis took all the following day.

In contrast, the Negev Brigade's attack on Iraq Suwaydan met stubborn Egyptian resistance and ended in failure. Having cleared the Palestinian villages north-west of Gat, the Givati troops arrived at their jump-off points facing the villages of Iraq Suwaidan, Bait Affa and Ibdis. However, the Negev Brigade troops had trouble reaching their positions because vehicles coming from Dorot, near Ruhamah, could get no closer than 5km south of Iraq Suwaidan. Nevertheless, this village was taken with relative ease. Not so the Iraq Suwaiden police fort. Here the Israeli attack began later than intended. The first barbed wire fence was breached by Negev Brigade troops, but the Egyptian defenders opened fire and although the next four fences were also breached, dawn came up while the sappers were still at work. Moreover, the gate had not been reached and so a withdrawal was ordered. As the Iraq Suweidan police fort had not been taken, Iraq Suweidan village now had to be abandoned.

At the end of these somewhat confusing battles on the Egyptian front, Israeli forces had gained the villages of al-Tinah, Bi'lin, Ibdis and Jalandiyah, followed by al-Masmiyah al-Saghirah (also known as Huraniyah). Meanwhile, in the north the first Israeli offensive began late on 8-9 July, and was a full-blown push to overrun what remained of the UN designated Arab areas of northern and central Palestine, particular attention being paid to Syrian forces defending eastern Galilee.

A Bren-gun team of the Egyptian Army in Palestine, operating their weapon. (Nour Bardai collection)

AIR OPERATIONS

On the Syrian front, the IDF/AF had intended to attack a forward Syrian Army HQ at al-Qunaitrah, but this was called off because no Dakota bomber was available. Other Israeli sorties were, however, carried out in the north; two of the Tel Aviv Squadron's Beechcraft Bonanzas and a Dragon Rapide dropping bombs indiscriminately on some enemy-held Palestinian villages in central Galilee in support of an Israeli Army attempt to conquer central Galilee and expel Arab Liberation Army units.

In the north, the Syrians and Lebanese seem to have been caught by surprise, and it took time for the SyAAF to respond. Nevertheless, the Syrians had been aware that things were about to heat up and in his diary entry of 8 July, Faisal Nassif recorded: 'I left Alice at her mother at five o clock on the afternoon. I left the airfield (al-Mizzah outside Damascus) at 6.30 in the evening and went to ---- (unclear). We left al-Dimas (between Damascus and Lebanese frontier) at 10.30 on our way to Istabl'. This last sentence is not very clear, as the following day would also involve travelling to Istabl.

Meanwhile in the Israeli POW camp, Abd al-Rahman Inan and his fellow prisoners had started their Ramadan fasting. In order to ensure sufficient for the evening meal which broke the daily fast, they decided to not to eat their already inadequate pre-dawn breakfasts but were nevertheless concerned that the food might go off during the July heat. On this, the first day of Ramadan, the three captured REAF pilots were delighted to each receive a parcel from Wing Commander Hashad who seems to have been attached to the Egyptian Mission in Amman. Presumably sent via the Red Cross, the parcels contained sweets, food and cigarettes. It was the first time they had heard from one of their colleagues. Unfortunately, the Israeli camp authorities seemed ignorant of the rules of Ramadan, and so presented the prisoners with their 'fast breaking' meal of boiled potatoes and a few eggs before the sun set, so the Egyptians had to just wait, looking at the meal, until the correct time to eat.

The Israeli offensive on the southern front would involve several operations by the Israeli Air Force during the night of 8-9 July. Bombers from the Tel Aviv Squadron were supposed to attack Bayt Jibrin, Iraq al-Manshiyah and Falujah, but only the first of these targets was found. According to some accounts the aircraft assigned to bomb Iraq al-Manshiyah hit the Israeli settlement at Ruhama instead. This seems unlikely, as Ruhamah is a full 14km to the south-south-west, while the crew assigned to attack Falujah, being unable to locate their target, jettisoned their bombs into the Mediterranean. Four Avias were also supposed to attack the REAF aerodrome at al-Arish, but were delayed by technical problems. They tried again during the morning, but one Avia S-199 crashed on take-off, while another was shot down by Egyptian groundfire near Gaza.

No information is available concerning REAF activity on 8 July and it is possible that nothing was flown from al-Arish on this day.

9 JULY 1948

GROUND OPERATIONS

On 9 July a number of explosions rocked parts of Cairo. Away to the east, in the Gaza area, the Egyptian Army continued to try to secure its still vulnerable and exposed lines of communication and supply. The 3rd Infantry Battalion, commanded by Colonel Abd al-Halim al-Dighaydi, achieved the second Egyptian success in this 'Ten Days War' when, at noon on 9 July, two infantry companies from the 3rd Battalion finally seized control of the Israeli settlement of Kfar Darom. This faced the Palestinian village of Dayr al-Balah, which had been besieged since 16 May. On the same day, the 9th Battalion, commanded by Colonel Kamal al-Rahmani, achieved another success north of the so-called Lateral Sector. Here al-Rahmani, having been ordered to seize the settlement of Negba, first captured Hill 113, which dominated the strategic Negev road junction. During the morning, two of his infantry companies, supported by two tank companies, succeeded in taking the hill from the Givati Brigade, meaning that the Egyptians could now use Hill 113 as a base from which to attack neighbouring Negba. Later during 9 July, these Egyptian troops retook Bayt Affa but not Ibdis. Overall, the Arabs had nevertheless lost al-Khayma, Barqusyah and Qastina on 9 July, followed by Tal al-Turmus the following night.

AIR OPERATIONS

Attempting to describe the events on the ground during the night of 8-9 July, and on 9 July itself, can appear confusing, as several of these operations continued through both periods. Events in the air are clearer in the REAF Operational Diary from al-Arish. Sortie No.1 took off at 0545 hours, almost exactly an hour after sunrise. It consisted of a single Spitfire 5, number 601, which was the first to be received by the Egyptians. It was flown by Wafa', whose task was an armed reconnaissance, almost certainly of Kfar Darom. Visibility was good and no enemy fire was observed. Following his return, at 0715 hours, Wafa' recorded that: 'Our artillery was observed shelling the settlement intensively and our vehicles were (seen) on the road from Gaza to Dayr al-Balah but no hostile movements were observed'. The Egyptian artillery barrage was in preparation for an assault by infantry by the 3rd Battalion, who took control of the Israeli settlement around noon.

A poor, but extremely rare photograph showing a pair of Egyptian Army-operated-, German-made Flak 18/36/37 anti-aircraft guns calibre 88mm – famous '88s' – as seen in action during the Palestine War. Although some of most effective artillery pieces of the time, widely used for fighting enemy armour and fortifications, and although number of Flaks were found on dumps left behind after the Second World War, acquiring ammunition for them was certainly not easy for Egypt. Together with the deployment of different other German-made tracked vehicles found at junkyards left behind following the campaign in the Western Desert of 1940-1943, this is indicative of how desperate to obtain additional armament the Arabs became during the summer of 1948. (David Nicolle collection)

The REAF Operational Diary makes no mention of Israeli fighters in this area; Sortie No.1 having taken off roughly an hour after IAF's Avia S-199s were supposed to attack al-Arish. But, nor did it mention Air Commodore Abd al-Mun'im al-Miqaati's brush with one of the Israeli Avias, because al-Miqaati's flight was not an operational sortie. In fact, four Avias of the IAF's No.101 Squadron had been allotted to attack al-Arish airfield, but one crashed during take-off, the pilot suffering minor head injuries, another crashed into the Mediterranean with the loss of its pilot, while trying to shoot down the Egyptian Air Commodore's Lysander (see below), and the others missed their target. Given the lack of success achieved by Tel Aviv Squadron bombers during the previous night, a military historian's description these sorties as 'a veritable festival of backfired missions' is probably justified. Apparently, the Israeli Army on the ground did not hesitate in telling the IDF/AF what they thought of the Air Force's performance.

Israeli accounts state that the three Avias which did take-off with their bomb loads, headed for al-Arish led by Lou Lenart. Reportedly unable to find the aerodrome because of heavy cloud, although according to the REAF Operational Diary visibility was good, they attacked what they identified as Egyptian military targets around Gaza and its 'port'. In fact, Gaza only had an open roadstead, not a port in the normal sense of the word. Two Avias, flown by Lenart and Al Freeman, returned to Herzliyya, but their colleague Bob Vickman (Weekman), an ex-USAF pilot with over 200 flying hours, 65 missions and many decorations, did not. He was assumed to have been brought down by Egyptian anti-aircraft fire, of which there had been plenty. This loss apparently had a profound impact upon the Israeli fighter pilots. Lou Lenart, the joint commander of No.101 Squadron, was transferred to the Air Headquarters, while Modi Alon became sole commander of what was as yet the Israeli Air Force's only fighter unit.

In fact, Bob Vickman had not fallen to ground fire but was lost when he attacked al-Miqaati's lumbering Lysander. Air Commodore al-Miqaati, the Deputy Director General of the REAF, had been making an inspection of the Tactical Air Force at al-Arish and had just taken off for his return to Cairo. The Egyptians had been expecting trouble and the aerodrome was at 'Condition Red' when al-Miqaati and his observer/air-gunner took to the air. Unfortunately, the serial number of al-Miqaati's aircraft is unknown, but it was probably a machine from the Communications Squadron rather than one of the reconnaissance Lysanders based at al-Arish.

Air Commodore Abd al-Mun'im al-Miqaati had been told to keep his radio on so that al-Arish could warn him of any danger. As al-Miqaati said in his interview with the author in 1973:

> I was still nervous as I set out across the Mediterranean. Fortunately my gunner — I don't remember his name — was a keen-sighted man and he spotted an Israeli Messerschmitt (Avia S199) as it manoeuvered into a position to attack. Of course, my Lysander was a very old kind of aeroplane but I'd flown these for a long time. Still, we were at a big disadvantage and you'd expect that such a contest could only end one way. The pilot of the Israeli aircraft came up behind us. I told my gunner to fire just before the Messerschmitt came into range and I then went down to about 100 feet (about 30m). Then the gunner fired and I throttled right back. You know the Lysander can drop like a stone to land in a field — like they did when the RAF took spies in and out of France (during the Second World War). The Israeli must have been concentrating on keeping me in his sights because he dropped his nose to follow me. He overshot and went right in, almost level with me. I honestly felt sick in my stomach and - I don't know why - I saluted him. Then we flew straight back to Cairo.

No wreckage of Bob Vickman's Avia was ever found, which suggests that the combat took place over the sea.

As soon as it became clear that the Israelis had launched a major offensive against the Egyptian Army, the REAF was instructed to strike back at Israeli airfields. The result would be claims of enemy aircraft destroyed on the ground, which the Israelis largely denied. The first to be attacked was Aqir (Tel Nof), which was targeted by REAF Sortie No.2, consisting of four Spitfire 9s carrying high-explosive and incendiary bombs. These Spitfires were flown by Abu Zaid, Galal, Yahya and al-Aidarus, in numbers 642, 663, 643 and 646 respectively. They all took off at 0810 hours and returned safely at 0925 hours. Visibility was good but enemy fire, described as Bofors, would be encountered. Their report described the sortie as: 'Bombing the airfield with bombs and gunfire. Bombs fell on a hanger which was directly hit. Also bombs fell on the tented camp, and some aircraft were hit on the ground by strafing with machine guns'.

A Bofors L/60 anti-aircraft gun calibre 40mm, mounted 'Portee'-style, on a Bedford truck, as seen during action in Palestine. (David Nicolle collection)

Three Spitfire LF.Mk 9s of the REAF, as seen lined-up at al-Arish in mid-1948. Notable is the application of the fuselage band on the aircraft in the centre left. (David Nicolle collection)

Reports that one or more Egyptian Spitfires also attacked Sde Dov aerodrome remain a mystery, as none of the Sorties recorded in the Operational Diary mention such a raid. This was probably carried out by one or more bombers operating from Almaza. The REAF certainly attacked Israeli concentrations in Negba and near Julis, as well as Tel Aviv harbour. In fact, the next operation to be listed in the al-Arish Operational Diary was Sortie No.3, which was both a bombing and a reconnaissance mission targeting Negba. It was carried out by pair of Spitfire 9s, armed as usual with HE and incendiary bombs. Number 651 was flown by al-Qarmuti, and number 657 by Muhammad Safwat. Both took off at 0850 hours, had good visibility, no noticeable hostile fire and landed back at al-Arish at 1010 hours, to report: 'Direct bombing of Negba settlement'.

Sortie No. 4 had the same purpose as the previous sortie, but was more eventful. It was carried out by Wahdan in Spitfire 5 number 615 or 610 (the serial number is unclear), and Tahir Zaki in Spitfire 5 number 616. They took off at 1005 hours, shortly before Sortie No.3 returned, and themselves returned at 1110 hours. Visibility had been good, no hostile fire was observed, and their report stated that: 'We hit groups of waggons inside a cultivated garden at coordinates 1202-1107 and destroyed some of them and we hit 2 waggons in the road between Julis and Nagba and we destroyed them'. Though the coordinates are written clearly in the Operational Diary, they do not make much sense. The lateral 1202 coordinate ran across the track leading into Negba settlement, but the north-south coordinate 1107 would put this location well to the south where there does not appear to have been a road or even a track. Vertical coordinate 1170

Refuelling of a Spitfire LF.Mk 9 already loaded with 50lb bomb installed on universal carrier under the left wing. (David Nicolle collection)

would, however, have placed the location close to the vital lateral road just west of Iraq Suwaidan village.

Sortie No.5 was again flown by a pair of Spitfire 9s, with the same purpose and carrying the same armament. Mustafa al-Utayfi in number 651 and Abd al-Azim in number 621 both took off at 1200 hours, reported good visibility and no hostile fire, landing at 1325 hours. They stated that: 'Three armoured personnel carriers were attacked next to the settlement (of Negba) and inside it, but nothing (else) was observed except three armoured personnel carriers to the east of Negba which were not attacked'.

Sortie No.6 would target Tel Aviv harbour (see below), but while this was still in the air another tactical support mission was launched, No.7, which flew over al-Hulaiqat looking for signs of the enemy. This time, a Spitfire 9 and a Spitfire 5 were involved, both armed with HE and incendiary bombs. Wafa' was in the Spitfire 9 number 636 or 646, while Wahdad was in the older Spitfire 5 number 616. They took off at 1400 hours and landed safely at 1525 hours, having experienced good visibility and no enemy fire, to report that: 'Negba settlement was directly attacked and six vehicles were bombed 1km to the east of al-Hulaiqat. There was no activity in settlements to the east of the road between Gaza and al-Majdal'.

On this notably busy day for the REAF, Sortie No.8 was a reconnaissance of Negba and Julis, plus close support for the Egyptian Army on the ground by attacking any enemy who were found. Thus Abu Zaid (who had returned from the attack on Tel Aviv harbour at 1350 hours) and Salim Tahir took off at 1730 hours in Spitfire 9s number 641 and 657 respectively. Salim Tahir would return at 1845 hours and Abu Zaid at 1900 hours. Visibility had been good, but they reported being fired at by Bofors guns. The comments column noted: 'Attacked groups of waggons and two trucks north-west of Negba by bombs and hit them. We hit enemy APCs on the Julis-Negba road and hit a truck in the same place. (Also) Hit truck in Be'er Tuvia by gunfire'.

President Gamal Abd al-Nasser's recollections of his time, defending what would become the 'Faluja Pocket', contains some interesting details which might refer to this date.

> During this time the Number 17 (unit) were preparing themselves to attack Negba for the second time. They asked team Number 6 to take the enemy position at Kilo 7 so that they (the enemy) could not get any ammunition through. At that time the only pilots available were Sq. Ldr. Abu Zaid (alone?) who was currently bombing the enemy in support of troops led by Naqib Mahmud Mahir al-Ramali, (now) the mayor of al-Minya at time of writing, who was also attacking on the ground.

Squadron Leader Muhammad Abu Zaid's earlier attack on Tel Aviv harbour was Sortie No.6, carried out in company with al-Qarmuti and Safwat, flying Spitfire 9s number 624, 641 or 643, and 636. They had taken off at 1230 hours and would return at 1350 hours, reporting good visibility and no hostile fire. The comments column merely states that: 'The port was attacked with HE and incendiaries which caused fires, and we also we hit ships in the port including a small ship about 200 tons which was hit'.

Israeli accounts of this raid were very different. A radio broadcast from Tel Aviv at 1630 hours GMT claimed that it took place at 1500 hours local time, dropping bombs indiscriminately, injuring 10 people, while one bomb hit 'educational establishments' and killed two 14-years-old girls. At 1830 hours GMT, the Voice of Israel claimed to quote 'an English transmission' which stated that only 6 bombs were dropped, which would be correct if only the HE bombs were counted.

This might also have been the raid witnessed by Abd al-Rahman Inan and his fellow prisoners. Seemingly there was a camp for displaced Palestinians not far away, and those interned here were seen and heard cheering the air raid. Israeli troops were accused not only of shooting into the air to quieten the crowd, but of firing several shots towards those involved, injuring some. When the Red Cross were asked to investigate, they were, in Inan's opinion, misled by the Israeli authorities, but without more accurate dates and locations, this story must be treated with caution.

The raid described by Inan may, in fact, not have been Egyptian at all. A Haganah statement, broadcast from Tel Aviv at 1630 hours,

A close-up look at a 50lb bomb (essentially a mortar bomb calibre 120mm) as installed under the left wing of a REAF Spitfire LF.Mk 9. The inscription reads: 'Guide me to Tel Aviv'. (David Nicolle Collection)

claimed that: 'Two enemy aircraft were reported to have attacked Jewish positions north of Jinin at 1030 (local time) on 9th July'. A broadcast from Jerusalem at 2000 hours GMT stated that: 'The Iraqi Air Force bombed Jewish-held areas near Jinin, while Egyptian aircraft raided Tel Aviv on 9th July and alerts were sounded in Haifa when unidentified aircraft, believed to be Egyptian, flew over the harbour without dropping bombs'. There is no evidence that the aircraft that flew harmlessly over Haifa were Egyptian, and this might have been a British reconnaissance sortie, although it seems more likely they were from the RIrAF.

Major Rashid Ali Awad certainly flew a combat reconnaissance in an Iraqi Anson over the port of Haifa, following intelligence reports that several ships were unloading contraband there. He found no targets and therefore reportedly dropped his bombs on Ramat David on the way back. This mission is said by unofficial sources to have been flown on 6 July, but seems more likely to have been on the 9th, while the bombs Major Awad supposedly aimed at Ramat David aerodrome may instead have been dropped in support of Iraqi units under Israeli pressure near Jinin. In fact, an Arab report broadcast from Jerusalem at 2000 hours GMT confirmed that: 'The Iraqi Air Force bombed Jewish-held areas near Jenin'.

On 9 July, Faisal Nassif of the SyAAF finally arrived at a front-line aerodrome, his diary recorded that: '9th July: We have left Damascus and have come to Istabl, (in?) Zarq (Lightning) Two. I have sent Alice to her parents. Sha'aban (a colleague, friend or relative) tells me that she has taken our belongings to her parents' home'. Lightning Two seems likely to be the name – official or otherwise – given to the Harvard aeroplane in which Nassif flew.

The urgency with which the SyAAF made its squadron at Istabl operational was justified, as the fragmented northern fronts were hotting up. On 9 July, at the expiration of the First UN Truce, Fawzi al-Quwaqji of the Arab Liberation Army had already launched an offensive, by sending his Hittin battalion from Nazareth toward al-Shajara, 14kms west of Tiberius. This village had fallen to Israeli, or as they had then been Zionist, forces back on 6 May 1948, but his troops made no real progress. The ALA attack was repeated several times during the following three following days, but was unable to dislodge the Golani troops who held this part of the Israeli front in Galilee.

Fawzi al-Quwaqji's men were also targeted by three Austers of the Israeli Air Force's Galilee squadron, which dropped 20/25kg bombs on four Palestinian villages. The first such raid was at 0555 hours against Tarshiha and Majd al-Krum, followed by al-Maghar 10 minutes later, and Shafa Amru at 0830 hours. Later in the day, the Israeli Austers dropped 9 anti-personnel bombs on Jaba' and Ain Razal (unidentified), while elsewhere small incendiary bombs and leaflets calling on the local Palestinians to surrender were scattered across other villages. According to a Haganah report, broadcast at 1730 hours GMT on the 9th: 'The large-scale Jewish operations carried out with the participation of infantry and aircraft, led to the occupation of the Arab village of Kuwaykat, 8km south-east of Nahariya'. This Palestinian village had held out, defeating two previous assaults earlier in the war, but its inhabitants were now expelled. During the evening of 9 July, Israel troops advanced eastward along the inland road from Acre to establish themselves, in cooperation with some Druze villagers, in hills to the east. There they waited to see the Lebanese response.

In the south the intended bombing of Innabah, 7km east of Ramlah, on the night of 9-10 July, by one of the Tel Aviv Squadron's Dragon Rapides was called off because the aircraft was not ready. Nevertheless, Innabah was taken the next day by the Israeli Army's Yiftah Brigade.

10 JULY 1948
GROUND OPERATIONS

Two significant Israeli offensives would be launched during the night of 9-10 July. That in the north was called Operation Brosh, started after nightfall on 9 July and would last until 14 July. One of

While Israeli-reported air raids by Arab air forces on the Jerusalem area were unsubstantiated, different parts of the city were regularly subjected to artillery shelling not only by the Israelis, but also by Transjordanians. (Albert Grandolini Collection)

Gunners of the Syrian artillery as seen during training, in 1949, in a scene reminiscent of their operations during the Palestine War. (Albert Grandolini Collection)

the Carmeli battalions managed to get across the upper Jordan river with some difficulty, but the second battalion was delayed and came under accurate Syrian artillery fire from both east and west. Word was then received that the Syrians were about to launch their own attack in the direction of Rosh Pina, 7km north-east of Safad. When it became obvious that the second battalion could not capture the frontier customs house before daylight, the Carmeli battalions were ordered to return to the western bank of the river.

At dawn, however, before having completed their withdrawal, the Carmelis were targeted by strong Syrian counterattacks on both sides of the Jordan. Suffering heavy losses, they were driven back to the western bank, where the other Carmeli battalions were fighting desperately to hold a Syrian advance that threatened Mahanayim and Rosh Pina, incurring similarly heavy casualties in the process. By noon, the Syrians had regained all their earlier possessions west of the river, and the Israelis had withdrawn to defend the approaches to Mahanayim and Rosh Pina. For two days, the Israelis desperately fought-off the Syrians to stem their advance, which nevertheless reached the settlements of kibbutz Hulata and Mahanayim. In this see-saw battle some positions changed hands four times.

Meanwhile on the central front, Israeli Prime Minister Ben Gurion had decided the first priority should be given to raising the Arab siege of Jewish West Jerusalem. The IDF High Command had therefore prepared an offensive called Operation Dani to secure the corridor to Jerusalem. The first phase was to occupy the towns of Lydda and Ramlah, pushing the front line as far to the east as possible to remove the threat to Tel Aviv, while seizing a major railway junction as well as the international airport. The second phase called for the capture of Latrun and Ramallah, to widen the Jerusalem Corridor and eliminate Arab pressure on the city of Jerusalem.

From the Israeli point of view, an attack on Lydda and Ramlah would compel the commanders of the Arab Legion to send part of their force currently at Latrun to support the defenders of those two towns. This would weaken the Latrun stronghold and enable Israeli forces to break through. About four brigades were allotted to the operation under command of Yigal Allon, with the Harel, Yiftach and 8th Armoured Brigade, plus the 2nd Kiryati battalion and the 3rd Alexandroni battalion in support. In addition, elements of the Etzioni Brigade were attached to the operation. These forces were to be supported by some artillery units equipped with 75mm and 65mm guns that had recently arrived from Europe, along with a significant effort by the rapidly growing IDF Air Force.

The plan of the first phase of Operation Dani called for a pincer movement by these forces. The 8th Armoured Brigade supported by two infantry battalions (the 2nd Kiryati and the 3rd Alexandroni), would swoop from the north, first to seize the international airport and then to drive into hill country north of Lydda to relieve Ben Shemen while also capturing several Palestinian villages on their way. Simultaneously, the Yiftach Brigade currently near al-Barriyah northwest of Latrun, would advance northwest through the hills, capturing further Palestinian villages on its way, before meeting the 8th Armoured Brigade at Ben Shemen. Once these two forces met, isolating Lydda and Ramlah, Yiftach and the 8th Armoured Brigade would together break into Lydda, leaving the other units to organize the defence of the villages previously captured. When Lydda fell, Ramlah would be isolated and could then be taken without difficulty. Meanwhile the Harel Brigade, which had been entrusted with the defence of the Jerusalem Corridor, would widen the southern part of this corridor and carry out diversionary operations towards Latrun.

At nightfall on 9 July, the Yiftach Brigade commenced the first phase of the operation. During the night and the following morning, it cleared the villages of Innabah, Jimzu and Daniyal. To the north, the 3rd Alexandroni battalion seized the German Christian settlement of Wilhelma and then the Palestinian village of Qulah, while the 2nd Tank Battalion of the 8th Brigade and the 2nd Kiryati battalion overran Lydda airport. Elsewhere, however, other units of the 8th Brigade encountered fierce Arab resistance. When it became obvious to the Arab Legion command that the Israelis had launched a major attack in the area of Lydda and Ramlah, the Legion's 1st Regiment, which had been holding the al-Nabi Samwil and Radar Hill, was hastily replaced by three less mobile garrison companies. The 1st Regiment then hurried north-west, to Bayt Nabala, in an attempt to halt the Israeli advance. As a consequence, the Israeli 8th Armoured Brigade not only encountered fierce local Arab resistance, but also faced Legionnaire counterattacks in the area of Qulah and Dayr Tarif.

The taking of Lydda aerodrome was nevertheless a significant gain for the Israelis, Prime Minister Ben Gurion apparently writing in his diary: 'The airport — who knows if the Israeli Government could have constructed such an airport in ten years – is in our hands. It is almost undamaged'. On the southern front, there were no further major developments on the ground during 10 July, though the efforts of the previous day were continued, including an Egyptian artillery barrage targeting Ibdis and supported by the REAF.

AIR OPERATIONS

Despite what some experts had forecast before the start of the Second Phase of Fighting, REAF Spitfires were able to maintain their pressure on the enemy for several days, attacking targets across a wide area and also helping the Arab Legion around Jerusalem. Thus, at the end of the 10 July, Egyptian Communiqué No.33 could claim with evident satisfaction that:

> Our bombers heavily bombed Tel Aviv airfield (Sde Dov), high explosive and incendiary bombs were used and several fires were started in various parts of the airfield, including grounded aircraft. Our aircraft also attacked Tel Aviv works and factories around the Auja river (Yarkon). Big explosions were observed and dense smoke rose from the factories. The Egyptian Air Force attacked the settlements of Negba, Be'er Tuvia, Dorot and (the abandoned Palestinian village of) Dayr Yasin. Heavy bombs and incendiary bombs were dropped and industrial instillations were destroyed. An oil dump received a direct hit and dense black smoke rose in the air.

The broadcast nevertheless failed to mention the loss of one aircraft.

The Tactical Air Force's Operational Diary included considerably more detail. Sortie No.1 was a straightforward raid, with HE and incendiary bombs, rather than an armed reconnaissance, targeting Negba and Julis. It was carried out by Spitfire 9s numbers 643 and 624, flown by Wafa' and al-Aidarus. They took off at 0455 hours and landed at 0625 hours, reporting good visibility, no hostile fire, and stated that: 'Attacked with HE bombs. Fire broke out in a building and intensive white smoke was observed rising into sky'.

Sortie No.2 did not go so well. This was a solo effort against al-Jammamah by Tahir Zaki in old Spitfire 5 number 601, armed with HE and incendiary bombs. He took off at 0500 hours but did not return – at least, not in his aeroplane. The remarks in the comments column in the Operational Diary were obviously written later, stating that: 'After attacking and destroying an enemy Auster plane, our pilot was hit by ground fire and was obliged to make a forced landing. The pilot has (subsequently) returned to his base safely (on foot)'.

Tahir Zaki did claim to have attacked an Israeli aircraft when interviewed by the author in 2001, but he seemed to be referring to a different day. It was not included in his painfully clear recollections of his own downing and subsequently escape. Unfortunately Tahir was by then of advanced age and sometimes conflated events which happened on different days.

To add further confusion, a report in the *Palestine Post* stated that an Israeli Auster Autocrat had been shot down near Lydda on 8 July. The bodies of the two crew were retrieved by the Arab Legion, handed over to the Red Cross and taken to Tel Aviv. But the REAF made no mention of air combat on the 8th. It might also be possible that the aircraft seen by Abu Zaid on the ground and strafed during Sortie No.4 was the machine already forced down by Tahir Zaki during Sortie No.2, though as Abu Zaid's primary target was Sde Dov aerodrome, this seems unlikely.

Tahir Zaki's escape to Egyptian lines would be reported in the Cairo daily newspaper *Al-Ahram* on 11th, along with a claim that an enemy Auster had been downed near al-Jammamah, though not necessarily by Tahir Zaki. By then, reports of Tahir's escape were taking on something of the character of a Wild West cowboy film. Egyptian communiqué Number 33, broadcast by Cairo radio at 2230 hours on the 10th, claimed that: 'All our aircraft except one returned

REAF pilot Tahir Zaki was lucky to escape with minor injuries when his Supermarine Spitfire Vb, serial number ER 601, was shot down on 10 July 1948. He then walked for many kilometres through Israeli-controlled territory to reach the Egyptian lines. (David Nicolle collection)

safely to their bases from all these operations. The aircraft which failed to return made a forced landing, the pilot returning safely after destroying his aircraft. He killed a number of enemy soldiers who pursued him and reached our lines'.

By the time Tahir Zaki turned up unexpectedly at an Egyptian Army outpost, his comrades at al-Arish had carried out an impressive number of sorties. The next two were against varied targets in the Tel Aviv area. Sortie No.3 was by a pair of Spitfire 9s carrying HE and incendiary bombs, flown by Yahiya in number 651 and Safwat in number 657. They took off at 0635 hours, presumably before Tahir Zaki was due to return, had good visibility and met no hostile fire, landing at 0750 hours to report that: 'The harbour warehouses area was attacked, white smoke was observed and some warehouses were completely destroyed'.

Sortie No.4 not only took off at the same time as Sortie No.3, but also landed at the same time. Presumably it was given a different number because it targeted Tel Aviv airfield (Sde Dov) and might have approached Tel Aviv from a different direction in order to confuse Israeli ground defences. As it happened, neither of these sorties were aware of being fired at. Sortie No.4 was also a solo effort by Squadron Leader Muhammad Abu Zaid, flying Spitfire 5 number 616. His report, as summarised in the Operational Diary's comments column, stated that: 'Hit aeroplanes in the airbase with concentrated strikes and the bombs fell in the middle of one group and destroyed them all, and fires were observed'.

Even before Sorties 3 and 4 returned, it was clear that Tahir Zaki was missing. So a combined search and ground-attack was launched, this being Sortie No.5. Lysander number 516 was flown by Ra'uf, and escorted by al-Utayfi, in a Spitfire 5 number 605. Whether the weapons mentioned, HE and incendiary bombs, were carried by both aircraft or only the Spitfire is unclear. They took off at 0735 hours tasked with looking for Tahir Zaki's aeroplane and also to hit Israeli positions at unidentified Bir Hayen Shedemo (perhaps Shdemot). On their return at 0840 hours, they reported good visibility and no hostile fire, but also that: 'The settlement was attacked and bombs fell on the target but the lost aeroplane was not seen'.

Before relating Tahir Zaki's own later version of his adventures, as given to the author, it is worth looking at an earlier account of what might have been the same incident. Its identity is not certain, however, because Tahir Zaki escaped from enemy-held territory more than once after being brought down by ground fire. It was published under the title 'Ants Nest' in the magazine *Al-Toubar* on 14 June 1998, and came from a senior Egyptian Army officer who was serving in Palestine at the time.

> One morning in 1948 in the command centre in al-Majdal town, the command centre in Gaza informed me by telephone that an Egyptian fighter plane had crashed near its lines and that it had successfully been able to rescue the pilot, Tahir Zaki. I asked them to send him immediately to al-Majdal. He arrived in less than half an hour. After enquiring I found out that he was on a solo reconnaissance mission of one of the valleys near Gaza (this was one of the khur or dry valleys which drained into the Wadi Hasi, which itself flowed into the sea between Gaza and al-Majdal). He was hit by enemy fire that forced an emergency landing. He managed to escaped (sic.) from the enemy patrol units that tried to capture him. He killed with his gun one of their soldiers. I immediately contacted al-Arish airport and told them of this information and asked them to send all the aircraft they could to hit these enemy gatherings. In the remaining sunlight hours (about seven hours) our fighter aircraft carried out continuous raids on these enemy gatherings within and outside the valley and caused heavy damage. The pilots' comments were that they had raided an ants' nest that scattered frantically, making an easy target. The ants were of course enemy forces.

King Faruk inspecting a bombed-up Westland Lysander at al-Arish, in summer 1948. (David Nicolle collection)

Tahir Zaki's own account, as related to the author, was more detailed and perhaps by then coloured by the passage of time:

> The second time I was hit was even more strange. General al-Muwawi, the Commander of the Egyptian Expeditionary Force, told us that the Israelis were setting up ambushes to cut the roads, and that they were using horses to move around because these were quieter than motor vehicles. Al-Muwawi wanted to know where these horses were stabled so he asked me to try and find out because by now I knew the area well. We usually flew a reconnaissance at six in the morning (he actually took off at 0540). This reconnaissance covered the area between the Egyptian and Jordanian lines, across the Negev to the Wadi Araba. That was when I saw two large buildings set at right angles to each other. While I was looking at them, I was shot at by a machine gun from the ground. This frightened the horses and I could see the legs of the panic-stricken animals sticking out of the buildings. So, I strafed the buildings and I think I killed all the horses. But my aircraft was also hit from the ground and the engine started to heat up badly, so I turned back.
>
> Three minutes later the engine became so hot that it was likely to catch fire, so I looked around for a place to make a belly landing. It was very hilly there but I selected a place between some hills, put the flaps down and descended. Then I saw that the ground was not suitable so I did a zoom to reduce speed. I also scraped the left wing along the ground to slow myself down, then I put my hands in front of my face. Three-quarters of the wing was torn off before the plane finally stopped.
>
> I got out and started wondering what to do. I knew I would be pursued because the people on the ground would have seen the stream of leaking glycol and would realize that I had to land somewhere. In fact, I was between the Egyptian and Jordanian Armies so I decided to make my way to Gaza. I remember I had a pistol, some water and some little tins of nuts (as he had on the first occasion he was brought down, or perhaps Tahir was confusing the two events). I shot (or shot at?) the first Israeli who came on the scene. He arrived on his own. I suppose he saw the crash and must have expected me to be dead. To put any other pursuers off the scent I went due north for five kilometres and I walked barefoot like a bedouin, hiding along the way. But I kept away from the bedouin themselves because some of them were helping the Israelis (the Banu Jubarat Tribal School would not have been far away). Then I turned west for twenty kilometres. I still avoided the bedouin even though I had no water.
>
> I was now heading towards the Egyptian positions between Gaza and al-Majdal. In July, at this time of year, there was ripe corn and this is what I ate while walking. Since I knew the area from the air, I knew that there were no big Israeli positions around, but on the third day (here Tahir Zaki recollections obviously go astray, as he returned to the Egyptian lines later the same day, as indicated by Egyptian communiqué Number 33, broadcast by Cairo radio at 2230 hours GMT on the 10 July) I saw a military position. I didn't know if it was Egyptian or Israeli, but there was a Palestinian with a small boy nearby, and they were going towards this army position. I stopped the man and pointed my gun at him. I asked him what army it was, and he said it was an Egyptian position. I was now feeling light headed and wondered whether to fight to the death or shoot myself. I told the Palestine that the boy must stay with me and that I would shoot both myself and the boy if he betrayed me. The Palestinian walked up to the position and came back with two soldiers.
>
> The officer in charge of the outpost was named Mas'ud Salah Zaki. We had been cadets together. He was very happy to see me because everyone had heard I was dead (or had at least failed to return). This was about 6 o'clock in the afternoon. They gave me water and something to eat. They also contacted General al-Muwawi with the news that I had turned up (the General must immediately have given the good news to al-Arish), and asked whether they should send me to him right away. He said no, wait until the morning. That was when a car arrived for me and took me to the HQ at al-Majdal. I gave General al-Muwawi my report and he told me that over the past three days there had been none of those ambushes by men on horses (here again Tahir Zaki's timing has clearly gone astray).
>
> Then he contacted Cairo and asked for a Dakota to be sent for me. I reached al-Arish about one o'clock in the afternoon and was flown home. But I had to go to hospital in Cairo first because I had broken four bones in my hand during the crash landing. They also found that I had minor fractures in both hips. They sent me home about six o'clock (pm) the same day. There I found my family was receiving condolences from various visitors because they believed I was dead. At first everyone thought I was a ghost or an ifrit (a spirit which takes the form of dead people), but then everyone celebrated. Lots of whisky was drunk, not only by my father but even by my sisters and other members of the family. Ten days later I was sent back to al-Arish.

Tahir Zaki went to the Heliopolis Sporting Club before returning to al-Arish and some of his friends still remember the terrible state of his feet after walking barefoot across the desert. It is worth noting that Tahir Zaki was not mentioned in the Operational Diary from the day of his downing until 18th. Given the state of his downed Spitfire number 601, of which Israeli photographs would be distributed widely, it is remarkable that he could walk at all.

The fact that the REAF continued to pound the Israeli settlement and airfield at Ruhamah, only 1.5km east of al-Jammamah, strongly suggests that Tahir Zaki had been able to inform al-Arish of this location's importance, either before his aircraft was hit or before he had to make his forced landing. Certainly, it would be attacked no less than seven times on this day alone. The first was Sortie No.6, in which Wahdan and al-Qarmuti took off at 0925 hours in Spitfire 5 number 616 and Spitfire 9 number 636. They had an easier time of it than Tahir Zaki, with good visibility and no apparent hostile fire. Following their return at 1025 hours, they reported that: 'Al-Jammama settlement (meaning the Israeli settlement of Ruhamah, east of al-Jammamah) was attacked and also four armoured vehicles were bombed to the north of Negba. All the bombs fell on their targets'.

The next REAF mission to target Ruhamah and/or al-Jammamah was Sortie No.11. Initially, Mustafa al-Utayfy and Salim (probably Salim Tahir) were sent to attack Dorot, Sawafir (there being three closely neighbouring Palestinian villages of this name) and Be'er Tuviya, with HE and incendiaries. They took off at 1425 hours in Spitfire 9 number 651 and Spitfire 5 number 616, returning at 1520 and 1530 hours respectively, to report good visibility and no hostile fire. However, the remarks column in the Operational Diary shows that their main effort was against al-Jammamah or the neighbouring settlement of Ruhamah, the unclear handwritten text in the available carbon copy stating that: 'Jammama (or Rumama) settlement was hit by bombs and they fell on the target, and about ten lorries were destroyed inside the colony and they were dispersed and most were hit. Hit big gun with its crew'.

This attack was followed up at 1510 hours, shortly before Sortie No.11 returned, by Sortie No.12, which consisted of a pair of Spitfire 9s. The pilots were al-Qarmuti in number 643 and Yahya (probably Yahya Husain) in number 646. On landing at 1615 and 1625 hours respectively, they again reported good visibility, no hostile fire, and stated that: 'We hit groups of lorries by bombs and guns and also 100 soldiers and most were hit. And also we hit a group of wagons which were covered by lids (hardtop trucks?) and some APCs and 2 were burnt and set on fire, and we hit 3 lorries that were outside the colony'.

Next came Sortie No.13, which focused upon the same target with HE and incendiary bombs. This time Muhammad Abu Zaid flew Spitfire 5 number 604, while his less experienced colleague Galal flew Spitfire 9 number 633. Again they had good visibility and faced no hostile fire, both taking off at 1527 hours and landing at 1625 and 1630 hours to report that: 'Hit groups of different waggons in the valley around colony and some were destroyed, including lorries full of equipment which burned. And we hit repair shops and an APC inside the colony with bombs, and other waggons were hit by guns and many soldiers hit around it (sic.)'.

Sortie No.15 again targeted al-Jammamah with the same weapons. Pilots al-Aidarus and al-Tarazi flew Spitfire 9s number 624 and 657, taking off at 1645 hours and landing at 1805 hours to report good visibility and no defensive fire. Their report merely stated that: 'Lorries and APCs around the colony were bombed and the bombs fell around them'.

The final mission of the day, Sortie No.16, was a further strike on al-Jammamah with HE and incendiary bombs. It was a solo effort by al-Qarmuti in Spitfire 9 number 651, taking off at 1710 hours and landing at 1815 hours, reporting continued good visibility and no hostile fire. The comments column stated: 'Hit hangers (perhaps tented hangers at the neighbouring airstrip) in colony and lorries, about nine were hidden under screens and hit groups of waggons with guns and some soldiers around the waggons'.

REAF Sorties Number 7, 8, 9 and 14 had different objectives. No.7 was launched at 0940 hours on a reconnaissance and bombing attack, with HE and incendiaries, aimed at Julis and Negba, a pair of Spitfire 9s (numbers 651 and 657) being flown by Wafa' and Galal. Like colleagues who attacked targets farther south, they had good visibility and met no hostile fire. Both landed at 1040 hours and simply reported that: 'All the bombs fell in their target area'.

Sortie No.8 was described as an armed reconnaissance, though the Operational Diary mentioned no specific target. It again involved a pair of Spitfire 9s, number 641, flown by al-Aidarus, and number 643, flown by al-Tarazi, taking off at 1110 hours, landing at 1245 hours, with good visibility and no hostile fire. This time the comments column was more informative, recording that: 'Nine armoured vehicles were bombed and two additional vehicles were attacked which were moving from Julis'. The Palestinian village of Julis had been overrun by the Israelis on 11 June, and its inhabitants evicted.

Sortie No.9 was specifically described as being in support of the Egyptian Army at Isdud, and conducted by a pair of Spitfire 9s being flown by Muhammad Abu Zaid, in number 646, and Safwat, in 657, both carried HE and incendiary bombs. The two pilots took to the air at 1300 hours, landing at 1425 and 1430 hours respectively, having enjoyed good visibility and seeing no defensive fire. They then reported: 'We saw about 100-150 vehicles, some of them big lorries and some APCs inside the settlement. We dropped bombs on them and most caught fire and there is also a fuel depot or oil store, and smoke rose from it very heavily, and we strafed it with gunfire'.

There was then a pause in attacks on these targets while the REAF focused on Ruhamah. Not until 1635 hours did the two Spitfires of Sortie No.14 take off and head for Tal al-Kabani and al-Fayli, neither of which have been identified. Wahdan in Spitfire 5 number 617 and Safwat in Spitfire 9 number 636 had good visibility, met no hostile fire and landed back at al-Arish at 1745 hours to report: 'Hit village by bombs and inhabitants ran off'.

The Operational Diary for 10 July was satisfied to summarise the day's activities as: 'All our planes returned safely except for one which made a forced landing. The pilot later returned safely'. Egyptian Communiqué Number 33 broadcast from Cairo later in the day, hailed these assorted raids as a success and also claimed the destruction of an enemy aircraft:

> Our aircraft discovered large concentrations of armoured vehicles and lorries scattered inside the settlement of al-Jammamah, in which there are ammunition and motor works, and attacked them several times during the day. More than 100 armoured vehicles were destroyed. During these operations, an enemy which attempted to intercept our aircraft was brought down in flames. Our aircraft protected our land forces during land operations and bombed enemy mechanized units.

The aeroplane which was shot down was presumably the Auster which Tahir Zaki claimed to have attacked, but certainly did not describe as 'an enemy which tried to intercept our aircraft', nor did he claim that it fell 'in flames'.

Given the intensity of REAF activity at al-Arish, it is unsurprising that other Egyptian pilots found themselves ferrying new machines to this front-line airbase. One such was newly qualified Pilot Officer Mustafa Wagdi Fadil. According to his logbook, on 10 July he flew Spitfire 5 number 617 from Hilwan to al-Arish. As Hilwan was the REAF's main maintenance centre for Spitfires, and indeed various other types, we might assume that serial number 617 was an overhauled machine heading for the front. According to the available pages of the Operational Diary, this aeroplane had yet to fly any operational missions. However, it would be sent on Sortie No.14 on the very day that it arrived. Wagdi Fadil would not be

A Spitfire Vb or Vc rolling for take-off from al-Arish airfield. By 1948, the version was considered obsolete. However, the REAF had no option but to continue flying them in combat, and aircraft with serial numbers 602, 607, 615, 616, 617, and 619 are known to have flown about two dozens of attack- and reconnaissance sorties in early July 1948 alone. (David Nicolle collection)

thrown into action so quickly, and did not appear in the Operational Diary until 22 July. On that day he flew the same Spitfire that he had ferried to the front. This was probably because, as a newly trained combat pilot, Wagdi needed to be familiarised with the situation and to get to know the area. Perhaps he had also been sent to al-Arish to replace the currently unavailable Tahir Zaki.

It seems that it was not until two days after the fighting had erupted into the Second Phase of the Palestine War that Syrian Air Force Harvards reappeared over the battlefield in support of Syrian, Lebanese and local Palestinian ground troops. According to an Israeli intelligence report: 'the SyAAF was active, particularly on the 10th, sending four or six aircraft very frequently and attacking our forces with bombs and machine gun fire'. Syrian Harvards certainly bombed the settlement of Mahanayim, 7km east of Safad, along with the nearby airstrip, in an attack coordinated with Syrian ground forces.

Some of these missions were flown from the Syrians' grass airstrip at Istabl in southern Lebanon, and one of the pilots who took part was Lieutenant Faisal Nassif on his first combat mission. Nevertheless, these Syrian sorties proved so effective that the Israeli Northern Front Headquarters demanded air cover from the Israeli Air Force. It arrived in the form of a pair of Avia S-199s from Herzliyya, their pilots soon encountering two Syrian Harvards patrolling at about 2,000m above Mishmar HaYarden.

The Israelis knew that the rear or observer positions in Syrian Harvards had been fitted with guns, so the Avias attacked from beneath, a former RAF pilot named Maurice Mann being credited with shooting down one which supposedly came down in Syrian-held territory near Mishmar HaYarden with the presumed death of its crew. Evidence from the Syrian side, however, shows that this Harvard did not crash and that its crew escaped unhurt. The second Harvard (serial number 206), flown by Sergeant al-Abid with First Mechanic Muhi al-Din Wadi as his observer-gunner, retreated across the frontier into Syrian airspace pursued by Avia number 107 flown by Lionel Bloch, a volunteer or mercenary pilot from South Africa.

Somewhere near al-Qunaitrah, the regional capital of Syria's Golan region, Bloch came in to finish off the Harvard whose gunner-observer, Muhi al-Din Wadi, may already have been wounded. But Muhi al-Din Wadi returned fire and hit the Israeli Avia, which crashed close to al-Qunaitrah, killing Bloch. The Harvard then returned to its own base, but Muhi al-Din Wadi subsequently died of his injuries. The pilot, Sergeant al-Abid, was from a family living in Hims, but which had originally come from what was now Saudi Arabia. Many years later, a younger member of this large and influential family operated SAM-6 anti-aircraft missiles against the Israelis, while another member of the family became a Lieutenant General in the Syrian Air Force.

A different but seemingly quasi-official version of this encounter almost certainly contains accurate details but probably muddles them up. This stated that SyAAF Harvard number 208 was flown by an Iraqi or Iraqi-trained pilot, named Captain Ibrahim or Salim Sam'an, and that this was the machine attacked by Lionel Bloch. It also gives a longer, and perhaps correct, version of the observer-gunner's name as Sergeant Muhi al-Din Atassi, and states that Bloch intercept the Syrian Harvard or Harvards over Ein Gev, in that sliver of Palestinian, now Israeli territory, on the south-eastern side of Lake Tiberius. In fact, Harvard number 208 was almost certainly the other SyAAF aircraft involved in this sortie.

Faisal Nassif's diary helps to clarify some of this conflicting information. His entry for 10 July stated that:

> We left Estabal at half past three in the morning in aircraft (number) 208 of Lt. Sam'an (flown by Sam'an with Nassif as observer-gunner, or Sam'an's aircraft flown by Nassif?). Our aircraft went into battle with enemy Messerschmitt (Avia S-199) aircraft. One of the enemy aircraft attacked aircraft (number) 206 of Na'ib al-Abid and Wahaba Tayar Muhi al-Din … (unclear) Wadi. The enemy aircraft was shot down. AI-rami (the gunner) was injured and died from his wounds. The name of the gunner was Muhi al-Din, the first mechanic who died. The enemy aircraft was equipped with three 20mm machine guns and two temporary (meaning unclear) 0.5 machine guns (in fact the Avia S-199 had 2 x 13mm machine-guns and 2 x 15mm cannon). (Then we carried out a) Bombing exercise.

Nassif made no mention of the Harvard supposedly shot down by Maurice Mann in the second Israeli Avia S-199, suggesting that it returned home safely. It may, in fact, have been Harvard number 208 in which Faisal Nassif himself was flying.

The official Syrian version of these events was included in wide-ranging Communiqué Number 29, broadcast from Damascus at 2150 hours, which made several inaccurate statements:

> The enemy forces were supported by aircraft … A squadron of our dive bombers attacked the settlement of Khirbat Yarda (Mishmar HaYarden, north-east of Safad) and destroyed most of it; our aircraft also destroyed important enemy instillations (sic.). Our Air Force made all-day attacks on the settlements of Najmat al-Subh (Ayelet HaShahar), Mahanayim and Rosh Pina, and Qatana airfield (unidentified), scoring direct hits. Our Air Force machine-gunned mechanised enemy units in the vicinity of Rosh Pina, inflicting losses. A squadron of our Air Force engaged in an air battle with enemy aircraft and shot one down after setting it on fire. The crew was burned. The rest of the enemy aircraft fled. Four heavy machine-guns and a 20mm gun were found in the destroyed aircraft which fell west of Qunaitrah. A squadron of our Air Force attacked enemy concentrations near Yesud HaMa'ala (north-east of Safad). The settlements of Atlil (unidentified) and Dardarah (probably al-Dhuhriya, in which case the target would have been the Israeli settlement of Ayelet HaShahar) were bombed and our aircraft left dense columns of smoke rising from them. One aircraft carried out reconnaissance flights along the front lines. All our aircraft returned safely to their bases.

According to Israeli sources, the pair of Syrian Harvards that attacked Mahanayim dropped just one bomb, which fell at a considerable distance from any relevant target. The Harvards that supported Syrian Army units near the Jisr Banat Ya'qub also enabled the Syrians to retake some positions on the frontier, but actually inside Syrian territory, which the Israelis had seized earlier. In fact, the activities of the SyAAF aircrew impressed their enemies, an Israeli intelligence report later stating that: 'The air activity was felt throughout the day; their aircraft dived bravely and harassed our forces constantly with bombs and machine gun fire. Several vehicles were damaged … I must point out, however, our carelessness in not camouflaging our vehicles against air attacks'.

At 1900 hours GMT on the 10th, a Lebanese broadcast managed to confuse the matter even further by stating that: 'Beirut radio reported that four enemy aircraft tried to raid Damascus on the morning of 10th July, but were intercepted and forced to jettison their bombs in remote places; a Jewish monoplane was brought down, its pilot being killed'. A report in the *Egyptian Gazette* added the incorrect detail that the Israeli aeroplane that crashed near Qunaitrah was 'of Italian make'. Quite where that idea came from is unknown.

For a while, the truth of Bloch's fate remained unknown to the Israelis, but then a blood-stained document fell into their hands, which told the story in some detail. This document was Lieutenant Faisal Nassif's diary, which was found on his body when he later fell in action over Galilee. As has so often been the case, the Israeli Air Force seemed unable to accept that one of their aircraft was shot down in air combat by Arab opponents, and instead suggested that the unfortunate Lionel Bloch shot off his own propeller.

During overnight battles of 9-10 July, in the north and centre, the Israeli Air Force reportedly failed to provide adequate support, usually because of the unavailability of aircraft and sometimes for technical reasons. Also, it is clear that greater IDF/AF effort was directed against the Egyptians than the Syrians. Nevertheless, the Voice of Israel did announce at 1045 hours GMT on the following day that: 'During the night of 9th-10th July, the Israel Air Force bombed al-Majdal and enemy positions around Iraq Suwaidan, and carried out combined operations with the Army, attacking enemy positions at Ramlah, Lydda and Bait Nabala'. The Israeli bombing raid against al-Majdal was confirmed the following day by Egyptian Communiqué Number 33, broadcast from Cairo, which reported that: 'Two enemy aircraft of the De Havilland (Rapide) type, attacked al-Majdal and dropped two bombs. One of them fell on the military hospital and five persons were injured. A representative of the International Red Cross was present'. On the following day the IDF/AF was quite active in the north where the Syrians and Iraqis as yet had nothing operational and available to compete with Israel's Avia S-199s.

11 JULY 1948
GROUND OPERATIONS

Because the Israelis took advantage of their positions to shell Jinin, the Iraqi 4th Brigade was committed to clear the heights of Mount Gilboa, overlooking that town, and to regain some lost Arab villages. So, on 10 July, after a short artillery bombardment, an infantry force of the 4th Brigade supported by a force of Palestinian irregulars mounted an attack on the village of Faqqu'a. After a long fight, this fell into Arab hands, followed by Muqaybilah. Meanwhile, the 4th Brigade commander committed another force to regain the Palestinian villages of Arranah, Dayr Gazal and Arabbunah, while a further force advanced to retake Jalamah and Sandalah. Meanwhile, Palestinian irregulars blocked the routes to isolate these villages. By the morning of 11 July, seven Arab villages had been retaken and the Israelis were driven back to the line of Mazar-Zar'in where they remained until the Second Truce.

On the central front, the advance units of the Yiftach Brigade had reached Ben Shemen, 4km east of Lydda, by nightfall on 10 July. But, to the north, units of the 8th Brigade were held up in front of Dayr Tarif. The next day, Major Moshe Dayan, the commander of the 89th Mechanised Commando Battalion of the 8th Armoured Brigade, bypassed Dayr Tarif and pressed forward towards Lydda, where units of the Yiftach Brigade were now encountering difficulties. Dayan's battalion, mounted on half-tracks and jeeps dashed into the town firing in all directions, breaking through the surprised local Arab militias in the outer defences before retracing its noisy path back through the town. This daring diversionary operation enabled the Yiftach Brigade to take advantage of the ensuing Palestinian panic to capture most of the town.

However, an Arab Legion company in the police station mounted a similarly daring counterattack, which, coupled with the appearance of an Arab Legion patrol consisting of armoured cars on reconnaissance, engaged Israeli units on the outskirts of Lydda. The inhabitants of the town and the Arab irregulars, who believed that this patrol was the van of an Arab Legion column coming to relieve the town, poured out of the houses and attacked Yiftach troops. After desperate hand-to-hand and house-to-house fighting, the defenders found that the Legion column had been an illusion, and Arab resistance rapidly weakened. Moreover, the Arab Legion company was itself ordered to withdraw from Lydda, while the armoured patrol was obliged to pull back for fear that the road back to its base would be cut. With the fall of Lydda came the Israeli seizure of the nearby airport, which was overwhelmed by tanks that the Israeli Army had somehow obtained from the British.

On the southern front, both sides were preparing themselves for a major battle, al-Masmiya al-Kabira having probably already been taken by the Israelis. Israel's Deputy Minister of Defence visited the Givati's Brigade's lines, while the Egyptian Army assembled an infantry battalion, an armoured battalion and an artillery regiment. As Israeli-held Bayt Affa threatened the Egyptian position in Iraq Suwaidan and Egyptian movements along the lateral road between al-Majdal and Falujah, Colonel Ahmad Tawfiq, the commander of the 2nd Battalion, was instructed to regain it. Two infantry companies, supported by two armoured vehicles, were assigned to this mission, during which Colonel Tawfiq intended to encircle and then overwhelm the Givati Force in Bayt Affa.

At 0900 hours on 11 July, the attacking force began its assault as planned and after a short struggle the Egyptians stormed the Israeli positions. When the Givati force attempted to withdraw to the neighboring settlement of Negba, they came under deadly fire from the armoured vehicles, which inflicted heavy losses. After capturing Bayt Affa, Colonel Tawfiq decided to exploit his success by pushing forward to seize Ibdis. At noon the same day, Egyptian troops assaulted this village, which fell into their hands after a brief resistance. Soon, the 2nd Battalion troops resumed their advance towards Ibdis heights to drive Givati troops from these hills as well. Noting that armoured vehicles could not climb the steep hills, Colonel Tawfiq instructed them to provide flanking fire while two infantry companies advanced under a smoke screen to attack Israeli position on the hilltops. However, as soon as the Egyptians approached the hills, they came under deadly Israeli fire, which forced them to retreat. The next day, two other attempts similarly failed to seize the Ibdis heights, so the attacking force was ordered to return to Bayt Affa and fortify the battered and now empty village.

AIR OPERATIONS

In the early hours of 11 July, an Israeli Dakota "bomber" finally managed to attack al-Qunaitrah on Syria's Golan plateau. There was reportedly a small but operational grass airstrip nearby, which the IDF/AF believed was being used by the SyAAF as a forward base or emergency landing ground. In fact, the area was found to be deserted, so the Dakota circled for a while. Its pilot subsequently reported that he had come under fire from 20mm anti-aircraft guns from a distance of around 15kms – which would seem entirely pointless – then dropped his bombs seemingly at random before returning home. Some are said to have fallen on or near Nazareth in central Galilee. For their part, the Syrians issued their Syrian Communiqué Number 30 at 2055 hours GMT on 11 July, reporting that: 'Last night a Zionist plane dropped some bombs on the region of Banat Ya'qub Bridge and al-Qunaitrah, which caused some damage and injury'.

A post-war photograph of a column of M4 Sherman medium tanks of the Egyptian Army, rolling down the streets of Cairo. Whether acquired from surplus stocks of the British Army before the Palestine War, or assembled from whatever the Egyptians could find on dumps left behind by the Second World War, the type remained the principal medium tank of the army until 1956. (Albert Grandolini Collection)

During the day there seems to have been a modest amount of air activity on both sides of the northern front. A reported four or six Syrian Harvards provided close support to their troops with bombs and gunfire in the morning; some returning in the afternoon to bomb the Israeli settlements of Mahanayim, Rosh-Pina and Ayelet HaShahar. In fact, these later attacks obliged some Israeli units to fall back a short distance. The SyAAF now appears to have felt more confident and according to the Syrian Communiqué Number 30, broadcast on the 11th:

> The Air Force concentrated its attack particularly on the settlement of Derderah (almost certainly Ayelet HaShahar, next to the abandoned Palestinian village of al-Dhuhriya), which was set on fire. After yesterday's defeat of the Zionist Air Force which lost one plane shot down and another damaged, they did not put in an appearance today (which was not the case) … All our planes arrived back from these operations safely.

According to the Voice of Israel, broadcasting at 1830 hours GMT: 'Israeli aircraft attacked in the western Nazareth area'. After nightfall, of course, the skies again belonged to the Israeli Air Force.

To the south, while Egyptian and Israeli units struggled for control of Bayt Affa and Ibdis close to the lateral road, the REAF directed most of its activity against Ruhamah, Dorot, Nir Am, Gvar'am, Aqir (Tal Nof), Be'erot Yitzhak, Be'eri, Be'er Tuvia, Sa'ad and Tel Aviv harbour. Official Israeli sources refer to a raid on Tel Aviv's Sde Dov aerodrome, and reported an attack upon Israeli forces near or within Jerusalem. A further Voice of Israel broadcast at 1830 hours GMT on the following day, referred to actions on the 10th and stated that: 'Arab aircraft attacked Jewish transports in the Negev'.

The prejudice displayed by some western war correspondence extended to all the combatants, but was most blatant concerning the Arabs. For example, an American CBS news broadcast at 1308 hours GMT, on 11 July, quoted its correspondent in Tel Aviv, John Donovan, who described the Arab air raids as 'hit-and-run operations of no consequence. US pilots flying with the Jewish Air Force are contemptuous of the Egyptians and promise to hit Cairo soon. This war is being won by the Jews at great odds, not because the Jews are so good, but because the Arabs are so bad'. Donovan's reference to the 'great odds' against the Israelis was clearly misleading and contributed to a David and Goliath Myth, which has persisted to this day.

For its part, Egyptian Communiqué Number 34 broadcast at 2230 hours GMT on the 11 July, offered a quite detailed analysis of REAF operations on a day which was regarded as notably successful:

> Egyptian bombers made a severe attack on Tel Aviv harbour, airfield and factories, which were subjected to heavy concentrated bombing with high explosive and incendiaries. Explosions and fires were observed in most of the bombed districts. Damage is expected to be heavy. Even at a considerable height, air crews felt the shock waves caused by the heavy explosions … Our heavy bombers (which normally meant Dakotas) attacked the Zionist settlements of Be'er Tuvia, Dorot, Negba, Be'erot-Yitzhak and Shdemot. Bombs registered direct hits on buildings and instillations. Several armoured cars and lorries inside the settlements were destroyed. Our fighters (normally meaning ground-attack Spitfires) attacked Dayr Yassin, Bayt Hakirim (unidentified but perhaps associated with the similarly unidentified airfield of Matar Bayt H-k-ma written on the side of a REAF Spitfire during the First Truce) and Bayit VeGan (recently the location of a British radar station) in the neighbourhood of Jerusalem. Direct hits were registered on the objectives. Enemy anti-aircraft defences in this region were machine-gunned by our fighters. Our fighters, while acting as air cover to Egyptian forces, observed an enemy armoured car concentration which was machine-gunned and bombed, a large number being destroyed and the remainder disabled. Our medium bombers (in fact, again, Spitfires) attacked Aqir airfield with heavy explosives and incendiaries, scoring hits on the objectives, setting fire to hangers and destroying two Messerschmitt (Avia S-199) aeroplanes. Three other aeroplanes were strafed. The Egyptian Air Force attacked Tel Aviv for a second time today, the power station south of Tel Aviv being bombed while fires caused by previous raids could be seen in houses and factories. Damage and destruction could be seen everywhere. All our planes returned safely to their bases.

The al-Arish Operational Diary contains the normal entries for 11 July 1948, plus two additional pages providing information about the weapons loads carried by various aircraft, which is more detailed than was given on any other date. This is very welcome to the military historian and might perhaps have been requested by REAF HQ in Cairo, but remains difficult to interpret because the some of the take-off times were omitted, along with all return times and the names of aircrew. Most of the figures in the sortie number column correlate directly with those in the ordinary Operational Diary page, but not all, strengthening the idea that these additional pages were for a different purpose.

Thus, Sortie No.1 was an armed reconnaissance and bombing mission targeting Shdemot. It involved Lysander number 516, piloted by Ra'uf, and Spitfire 5 number 615, piloted by al-Utayfi. Both took off at 0455 hours and landed exactly an hour later, having had good visibility and meeting no hostile fire. The pilots reported that: 'Bombs fell inside the settlement accurately and 50 vehicles, including armoured vehicles (APCs?) were observed scattered to the north of the settlement. We think they were hit yesterday'.

The additional page indicated that Sortie No.1 had been an armed reconnaissance aimed at Shdemot, repeating that 10 armoured lorries had been destroyed inside the settlement, but updating the number of enemy vehicles observed outside the colony to 90 APCs and lorries. One serial number was also added, though without stating the aircraft type. Furthermore, this number was clearly written as 455, which would have been a Harvard, and the aircraft was stated to have been carrying HE and incendiary bombs. The additional page then went on to list the participating Lysander's weapons load as 4 x 250lb bombs, which would seem to be beyond the type's normal load, while that of the Spitfire was 12 x 25lb incendiary bombs.

According to the main entry in the al-Arish Operational Diary, Sortie No.2 was a solo armed reconnaissance of the Israeli settlements of Gat and Be'er Tuvia by pilot al-Aidarus in Spitfire 9 number 651, carrying HE and incendiary bombs. He took off at 0515 hours, found good visibility and no hostile fire, landing at 0655 hours to report that: 'Bombs fell in the settlement and directly struck the buildings'. The additional pages added that the target bombed was Be'er Tovia, and that the Spitfire had carried 2 x 250lb HE bombs, plus 4 x 25lb incendiaries.

Sortie No.3 was an ambitious affair involving six aircraft attacking several targets in and around Tel Aviv. Because the Israelis now had Avia S-199 fighters, some of which were flown by experienced if somewhat experienced, largely Jewish but non-Israeli, veterans of the Second World War, the REAF decided that its unarmed Dakotas bombers should have a fighter escort in the form of two Spitfires and the Hawker Fury prototype. The latter were also expected to bomb enemy targets. The Hawker Fury was probably carrying out its first operational mission in its new REAF markings and serial number and, like the Spitfires, carried bombs.

The main entry in the Operational Diary records that two Dakotas, number 803, crewed by al-Araja and al-Jamal (perhaps Jamal al-Afifi), and 806, crewed by Kafafi and Hamdi, took off first at 0930 hours, closely followed by Abu Zaid in the Fury at 0935 hours. Five minutes later, Safwat in Spitfire 9 number 657 took to the air, being joined by Wahdan in Spitfire 9 number 646, and al-Utaify in Spitfire 5 number 615, at 0945 hours. They all landed back at al-Arish at 1100 hours, reporting good visibility, no hostile reaction, and stating that: 'Spitfires are protections for Fury and Dakotas. The port and airport were bombed intensively and (enemy) losses were huge. The factory area to the south of the Auja river was attacked with 500lb bombs which caused severe destruction to buildings. The pilot felt shock-waves from the bombing at an altitude of 5,000 feet (over 1,500m)'.

The page of additional information noted that Dakota number 806 carried 10 x 500lb HE bombs, the Fury 8 x 250lb bombs, and the three Spitfires 16 x 25lb anti-personel and incendiary bombs. Tahir Zaki was still in hospital at this time, but he recalled that his colleague Muhammad Safwat bombed a munitions factory in the north of Tel Aviv that was providing the Israeli armed forces with assorted supplies. He claimed that this factory had become a source of concern to the Egyptian High Command and the Egyptians knew it by a name something like 'Btrir'.

However, Egyptian Communiqué Number 35, broadcast the following day at 2230 hours GMT, gave an incorrect time of the attacks on factories, unless there was another raid by aircraft operating from one of the Cairo aerodromes, and makes no mention of a munitions factory. It merely stated that: 'The Egyptian Air Force's heavy bombers last night raided Tel Aviv, dropping high explosives and incendiaries in the target area, hitting iron factories, the Ford car factories, water reservoirs and pumping stations; many fires were started and much damage was done'.

For their part, the Israelis identified the attackers as a C-46 Commando or C-47 Dakota escorted by Spitfires, thus opening up the possibility of a long-range night raid from Cairo by a REAF C-46. The recollections of Les Shegam, who flew an Israeli Avia S-199, might be more accurate, at least where the Egyptian aircraft were concerned. He was apparently scrambled in Avia number D-108 to intercept two Dakotas and two Spitfires, but saw no enemy.

This might also have been Adli Kafafi's first bombing mission in a Dakota from al-Arish, as he had previously been flying transport missions based at Almaza. Hassan Tawfiq's daughter Mona, who was 11 years old at the time, recalled that Dakota aircrew including Jamal al-Afifi, who was perhaps part of the crew of Dakota number 803 on this raid against Tel Aviv, used to come to the Tawfiq family home to relax after night bombing sorties from Almaza, though not of course when operating from al-Arish.

Sortie No.4 was flown in support of the Arab Legion troops, Palestinian militias, and the Egyptian Light Force, which were under pressure west of Jerusalem. It would end up controversial and a subject of competing propaganda claims. The most reliable account is the al-Arish Operational Diary, though it only provides an outline. In fact, Sortie No.4 targeted Dayr Yassin, the site of the most notorious massacre carried out by Israeli forces on 9 April 1948, along with Bayit VeGan and Jakarm, which is unidentified. It was carried out by three Spitfire 9s carrying HE and incendiary bombs: number 633, flown by al-Qarmuti, number 636, flown by al-Tarazi, and number 643, flown by Yahya. All took off at 1015 hours, returning at 1145 hours after experiencing good visibility and no hostile fire, to report simply that: 'Bombs fell exactly on the targets'. However, the page of additional information added a Lysander to the flight, though its number and pilot are not given, which had taken off 10 minutes earlier. This Lysander was stated to be carrying 2 x 250lb HE bombs, while the Spitfires carried 10 x 25lb incendiaries. According to this additional page, Dayr Yassin and Bayit VeGan were strafed as well as bombed.

This was probably the raid condemned in a Haganah report, repeated from Tel Aviv at 1730 hours on the 12th, which claimed that:

> Leaders of all religious sects in Jerusalem were to ask for the air bombardment of Jerusalem to cease, but, said Haganah radio, there was no hope of success. The first Egyptian air attack was said to have taken place on 11th July, on western Jerusalem, where four Jewish children were killed, and this criminal Egyptian attack was expected to be the beginning of a series.

Another version of the story maintained that as three Egyptian Spitfires approached Jerusalem they were fired upon by anti-aircraft gunners, believed to be Iraqi, stationed at Biddu north-west of the

Reportedly taken at Cyprus after the Palestine War, this photograph shows the REAF C-47 Dakota serial number 816, wearing a camouflage pattern introduced during the conflict (see colour section for details). (David Nicolle collection)

A pilot and a member of the ground crew jointly loading 50lb bombs on one of REAF Spitfire LF.Mk 9s. (David Nicolle collection)

Pilot embarking the Spitfire LF.Mk 9 serial number 643, in early July 1948, at al-Arish airfield. Notably, by the time, the aircraft not only retained the 'complete scoreboard' of Egyptian Spitfires in front of the cockpit, but received its complete national marking (and, invisible on this still: its serial on the rear fuselage, as well as the fin flash). (David Nicolle collection)

city. Two of the Egyptian fighter-bombers jettisoned their bombs and flew home, while the third supposedly dropped two bombs on the (or perhaps 'a') Jewish Quarter, killing three children and injuring a fourth. This supposedly happened at 1300 hours, which, if referring to local time, would be when Sortie No.5, not Sortie No.4, was in the air. If the time was GMT then Sortie No.6 might be the subject (see below). Three days later a broadcast from Damascus, repeating a news bulletin by the British radio mouthpiece 'Sharq al-Adna', which was itself quoting an unspecified radio report from Jerusalem, stated that: 'the Jewish airfield constructed between these two places (Romema and Giv'at Shaul in western Jerusalem) had been set on fire'. Whether any airfield existed in such a rugged location remains unclear.

It may be significant that the pages of additional information in the al-Arish Operational Diary expressed no interest in Sorties Number 5 to 8. Perhaps the REAF HQ in Cairo needed detailed information about the earlier sorties in order to counter claims by their enemy that the REAF had behaved in an illegal manner.

According to the REAF Operational Diary, Sortie No.5 was an armed reconnaissance and bombing mission with no specified target, involving Spitfire 5 number 604, flown by al-Aidarus, and Spitfire 9 number 624, flown by Mukhtar Sa'id, both carrying HE and incendiary bombs. They took off at 1245 hours, had good visibility and faced no hostile fire, landing again at 1415 hours to report that: 'Reconnaissance of Hamamah and Barbara villages. Bombing of Gvar'am settlement'. These three localities lay between Gaza and Isdud, nowhere near Jerusalem.

Sortie No.6 was described as 'open attacks' – perhaps what might otherwise be called free hunting. It consisted of a pair of Spitfires armed with HE and incendiary bombs, Spitfire 5 number 615 being flown by Wahdan, and Spitfire 9 number 646 by al-Arusi (Khalil Jamal al-Din al-Arusi). They took off at 1610 hours and returned at 1710 hours, having good visibility but being subjected to anti-aircraft fire, which they identified as Bofors. Their report stated that: 'Wart Settlement (presumably Kfar Warburg) was attacked, fires were ignited and some buildings were destroyed'. If the identification of the target as Kfar Warfurg is correct, then it, like the targets of Sortie No.5, was in the coastal sector, nowhere near Jerusalem.

Sortie No.7 was an attack on the IDF/AF at Aqir aerodrome by two of the Tactical Air Force's most experienced pilots. Muhammad Abu Zaid flew the old Spitfire 5 number 617, and Jalal flew Spitfire 9 number 636, both carrying HE and incendiary bombs. They took off at 1600 hours, had good visibility and no apparent hostile fire (though that sounds unlikely given their target), and landed safely at

REAF Spitfire-pilot *Qa'id Surb* Mukhtar Said flew a significant number of combat sorties during the early- and middle parts of the Palestine War. He would be killed in action on 21 October 1948. (David Nicolle collection)

1725 hours. Their report stated that: 'Aqir aerodrome was attacked, fires were ignited in the aircraft hangars and three aeroplanes were destroyed in the dispersals (shelters?)'.

The REAF's last sortie of the day was No.8, which again targeted Sde Dov aerodrome. It was carried out by al-Qarmuti in Spitfire 9 number 641 and Yayha Husayn in Spitfire 5 number 607, both, as usual, carrying HE and incendiary bombs. They took off at 1615 hours and landed back at al-Arish at 1730 hours. Having good visibility and facing no apparent hostile fire, they reported: 'The electricity generating station was bombed and several small ships were observed carrying small loads'. The generating station was probably the large Reading Power Station overlooking Sde Dov airfield. According to the Voice of Israel: 'Enemy aircraft appeared over Tel Aviv at 1900 on the 11th, for the third time that day. Seven persons had been injured, two of them seriously, when three enemy aircraft, including two twin-engined, bombed Tel Aviv shortly before 1300'. The Egyptian air raid, which included two Dakotas, had, in fact, taken place over an hour earlier, even taking into account local time differences. The impact of these raids was such that even at 1045 hours GMT on the following Tuesday, a Voice of Israel broadcast stated that what is called 'the Egyptian terror bombardment on Sunday had been the fiercest ever'.

The second round of fighting in the Palestine War of 1948 was now in full swing – and a lot more was to follow. That story is going to be continued in Volume 12.

BIBLIOGRAPHY

BOOKS

Abu Sitah & Salman Husayn, *Atlas of Palestine 1948* (London 2004).

Ali, F.K., *Al-harb al-'Arabia al-Israilia 1948-1949,* in Arabic [The Arab Israeli War 1948- 1949] (Beirut 1982).

Anon., *Tarikh al-Quwat al-Iraqiyat al-Musalat. Al-Jazz' al-Saba 'Ashar. Ta'sus al-Quwwat al-Jawiyah wa Tatwirha*, in Arabic [History of the Iraqi Armed Forces, Part 10. Establishment and Development of the Air Force] (Baghdad 1988).

Arena, N., I *Caccia della serie 5, parte Seconda. Fiat G.55* (Modena 1976).

Arena, N., I *Caccia della serie 5, parte Terza. Macchi 205* (Modena 1976).

Baghdadi, Abd al-Latif al-, *Mudhakirat*, in Arabic [Recollections] (Cairo 1977).

Be'eri, E., *Army Officers in Arab Politics and Society* (New York 1970).

Cooper, T., & M. Sipos, *Wings of Iraq: Iraqi Air Force, 1931-2003, Volume 1* (Warwick 2020).

De Gayffier-Bonneville, A-C., *L'echec de la monarchie egyptienne, 1942-1952* (Cairo 2010).

Goldschmidt, A., A.J. Johnson & B.A. Salmoni, *Re-Envisioning Egypt 1919-1952* (Cairo 2000).

Halim, N. Abbas, *Diaries of an Egyptian Princess* (Cairo 2009).

Hamlin, J.F., *The de Havilland Dragon Dragon/Rapide Family* (Tunbridge Wells 2003).

Ilan, A., *The Origins of the Arab-Israeli Arms Race. Arms, Embargo, Military Power and Decision in the 1948 Palestine War* (New York 1996).

Imam, Abdullah, *Wagih Abaza, Sahafat min al-Nidal al-Watani,* in Arabic [Wagih Abaza, Pages from the Nationalist Struggle] (Cairo 1995).

Inan, Abd al-Rahman, *Kanat Isiran*, in Arabic [I was a Prisoner] (Cairo 1971).

Jabr, Jabr 'Ali [Gabr, Gabr Ali], *Al-Quwwat al-Jawiyah bayn al-Siyasat al-Misriyah wa'l-Isra'iliyah, al-Jazz' al-'Awal 1922-1952,* in Arabic [The Air Force between Egyptian and Israeli Policies, volume 1, 1922-1952] (Cairo 1993).

Kafafi, M., *Fi Dikhran 15 Mayu 1948. 'Andama Istushihdan Abi! 'Azaf 'alay 'Uwtar al-Dhikra,* in Arabic [In Memory 15 May 1948. When my Father was Martyred! A Tune upon the Strings of Memory] (Cairo 1975).

Labib, Ali Muhammad, *Al-Quwat al-Thalāthah,* in Arabic [The Third Force] (Cairo 1977).

Lee, D., *Wings in the Sun: A History of the Royal Air Force in the Mediterranean 1945-1986* (London 1989).

Marr, D.S.B., *A History of 208 Squadron* (Southend 1966).

Marszalkiewicz, J., *Air Combat During Arab-Israeli Wars* (Lublin 2017).

McGregor, A., *A Military History of Modern Egypt, From the Ottoman Conquest to the Ramadan War* (Westport 2006).

Nomis, L. [with B. Cull], *The Desert Hawks. An American Volunteer Pilot's Story of Israel's War of Independence, 1948* (London 1998).

Nordeen, L., & D. Nicolle, *Phoenix over the Nile. A History of Egyptian Air Power 1932-1994* (Washington 1996).

Raafat, S.W., *Privileged for Three Centuries. The House of Chamsi Pasha* (Cairo 2011).

Sayigh, Y., *Arab Military Industry, capability, performance and impact* (London 1992).

Shakib, I., *Harb filistine, 1948, Ru'iya misriyya,* in Arabic [The Palestine War, 1948: An Egyptian Perspective] (Cairo 1986).

Sipos, M., & T. Cooper, *Wings of Iraq, Volume 1: The Iraqi Air Force, 1931-1970* (Warwick 2020).

Smisek, M., *Czechoslovak Arms Exports, Volume 1: Origins, Israel and Jordan 1948-1989* (Warwick 2021).

Stuart-Paul, R., *The Royal Saudi Air Force. A Legacy of His Majesty King Abdul Aziz Ibn Saud* (London 2001).

Sturtivant, R., *The Anson File* (Tonbridge 1988).

Vatikiotis, J., *Nasser and his Generation* (London 1978).

ARTICLES AND ANTHOLOGY CHAPTERS

Andersson, L., 'Wings Over the Desert: Aviation on the Arabian Peninsula. Part One Saudi Arabia', *Air Enthusiast, 112* (July-August 2004) 39-43.

Anon., 'How did you join the Air Force?' [recollections of Air General Mahmud Sidqi], *Al-Quwat al-Musalah* (November 1959) 22-23.

Anon., 'Pilot Shakib's Accident', *Al-Tubar* (14 June 1998) translation provided by Group 73 Aviation Historians, Cairo, page numbers unknown.

Anon., 'Success story in the sky' [REAF bomber operations in the 1948 War & the experiences of Captain Yussuf Fahmi], cutting of published article, date unknown; translation supplied by Group 73 Aviation Historians, Cairo.

Badr, Capt. Salah, 'Diary of Capt. Salah Badr' [commander of Egyptian Army garrison, Iraq Suwaydan, 1948] translated into Hebrew, *Bamahane* (8 November 1960) 11-13.

Baghdadi, Abd al-Latif al-, 'Ma qabl al-dubbat al-ahrar [Before the Free Officers]', *Hadhihi al-Thawra* (Cairo 1953) 188-9.

Biyali, Muhammad al-, 'At the REAF Operational HQ' (in Arabic), *Akher Sa'a* (8 September 1948) translation supplied by Group 73 Aviation Historians, Cairo, page numbers unknown.

Biyali, Muhammad al-, 'Okra Sandwich' [account of a bombing raid on Tel Aviv by REAF Dakotas] (in Arabic), *Akher Sa'a* (8 September 1948) translation supplied by Group 73 Aviation Historians, Cairo, page numbers unknown.

Deraz, Issam, 'Mudhakirat 'an 'Ida' al-l-wa'l-tayar: Tahir Zaki Hassan Zaki [Recollections of Taher Zaki Hassan Zaki]'. *Al-Manar al-Jadid* (date unknown) copy of the author's pre-publication typescript provided by Tahir Zaki, 2005.

Mahbub, Ali Muhammad (Lt.), 'Conditions of Flexibility in the Tactical Air Force', *Mujallat al-Silah al-Jawi al-Malki [RAF (Egyptian) Air Force Magazine],* 10 (1948) 17-23.

Siddiq, Muhammad Kamil (Major), 'Air War in Palestine' [in Arabic], *Mujallat al-Silaj al-Jawi al-Malki* [*RAF (Egyptian) Air Force Magazine*], 10 (1948) 29-31.

Taha, Sayyid, Brig., (diary extracts from 1948, published in 1950) photographed but unattributed cutting supplied by Brig. Dr. Abd al-Rahman Zaki, 1971.

Zaki, Tahir, 'The pilot who attended his own funeral', *Al-Tubar* (21 June 1998) cutting supplied by Group 73 Aviation Historians, Cairo; page numbers unknown.

INTERVIEWS & PRIVATE CORRESPONDENCE

Abu Zeid, S.S., Interviewed by the author concerning his uncle, Muhammad Abd al-Hamid Abu Zaid (Alexandria 2000).

Cooper, T., Correspondence concerning the Royal Iraqi Air Force from the 1930s to the 1950s (1997 to 2015).

Gazerine, I.H., Group Capt., Correspondence & follow-up information on the REAF following an interview with the author (1973-1974).

Hinnawi, M.S. al-, Air Vice Marshal, Interviewed by the author (Cairo 1999).

Kafafi, M., Interviewed by the author concerning her father, Adli Kafafi (Cairo 1999).

Miqaati, M. Abd al-Muna'im al-, Air Comm., Interviewed by the author (Cairo 1973).

Razzaq, A. Abd al-, Brig. Gen., Interviewed by Tom Cooper (supplied to the author 2007).

Sabri, F., Interviewed by the author concerning members of his family in the REAF & EAF (Cairo 2011).

Shafei, I. el-, Interviewed by the author concerning his father Adli al-Shafa'i (Cairo 2011).

Sidqi, Mahmud, Air Marshal [Muhammad Sidqi Mahmud al-Milaigy], Anonymous interview in the late 1950s (text supplied by Nour Bardai).

Sulayman, Abd al-Hamid [Suleiman, Abdul Hamid], Interviewed in Cairo by Gabr Ali Gabr 1988 (transcript supplied to the author 2012).

Tawfiq, M. [Mona Tewfiq]. Interviewed by the author concerning her father Hassan Tawfiq (London 1997 & 1999).

Zaki, Kamal, Interviewed by the author (Cairo 1999).

Zaki, Tahir, Interviewed by the author (Cairo 1997 & 1999).

NATIONAL ARCHIVES (LONDON)

AIR 2/4810 – Loan of RAF personnel & equipment to the RIraqAF, 1940-1950.

DEFE 4/16/145 – Report on the RIqAF, 11 October 1948.

DEFE 5/8/115 – Report on the RIqAF, 7 October 1948.

FO 371/69188 – Confidential report on current strength of REAF from British Air Attache in Cairo, 19 October 1948.

FO 371/69223 – Foreign Office on Ramat David and other incidents involving British military personnel from British Embassy in Cairo, 14 June 1948; letter of British Ambassador's meeting with the Egyptian Prime Minister, 14 June 1948.

ONLINE SOURCES

Aboul-Enein, T., 'The Writings of Abdullah Imam on Flight Commander Mohammed Wagih Abaza', *Small Wars Journal,* 845 (26 August 2011): https://smallwarsjournal.com.

Andersson, L., *Wings Over the Desert, Aviation on the Arabian Peninsula, Part 2* : http://www.artiklar.z-bok.se/Arabia-1.html.

Morris, B., *Arab-Israeli War* [on war-crimes during the Arab-Israeli wars]: www.crimesofwar.org/a-z-guide/arab-israel-war.

UNPUBLISHED SOURCES

Anon., *Biographical Note on the late Squadron Commander Mohamed Abdul Hamid Abu Zeid* (EAF Historical Department, n.d., supplied to the author 2001).

Anon., *Bab al-Sadis*, Chapters 29-30: 'Tactical Air Force operations', 421-37; 'Officers who Fell', 504 (Egyptian Ministry of Defence restricted circulation study of operations during the Palestine War, n.d.).

Anon., *Outline of the Career of Qa'id Liwa Jawi [Military Staff] Muhammad Hassan al-Maghribi* (Egyptian Ministry of Defense, unpublished typescript, n.d.).

Anon., *Struggle and Championship, The late pilot engineer Abdel Hamid Mahmoud [Abd al-Hamid Mahmud]* (Egyptian Ministry of Defense, unpublished typescript, n.d.).

Anon., *The Faluja Battle* (Egyptian Ministry of Defense, unpublished typescript, n.d.).

BBC Monitoring Service, *Trans-scripts of Middle Eastern Broadcasts during the Palestine War 1948-49* (BBC Monitoring Service, Caversham).

Fadil, Mustafa Wagdi, *Pilot's log books 1948-49* (selected pages supplied to the author).

Foschini, E. *Reports for on the supply, use and maintenance of Macchi MC205 aircraft to the REAF 1948-49* (written for the Macchi Company; supplied to the author by G. Cattaneo, 2007).

Gabr, Gabr Ali, Air Brig. Dr., *The Arab Israeli Conflict: The Roots and the Wars 1897-1979: An Egyptian Perspective* (unpublished book text, 2007).

Hartoch, N., *A History of the Syrian Air force 1947-1967* (PhD thesis, London University 2015).

Hinnawi, Shalabi al-, 'Eagles of the Air in Palestine' [on the 1948-49 Palestine War, in Arabic] (unpublished book chapter).

Nadif [Nassif], Faisal, *Diary*, recovered from his body on 16th July 1948 (English translation of Hebrew translation of original Arabic text; supplied by the IDF, 29 November 1995).

Stafrace, C., *The Iraqi Air Force* (pre-publication draft supplied by the author, 2008).

Various., *Operational Diary of the REAF Tactical Air Force [1948-49]* (incomplete photocopy).

ABOUT THE AUTHORS

DR. DAVID C. NICOLLE

David Nicolle is a British historian specialising in the military history of the Middle Ages, with special interest in the Middle East and Arab countries. After working for BBC Arabic Service, he obtained his MA at SOAS, University of London, followed by a PhD at the University of Edinburgh. He then lectured in art history at Yarmouk University in Irbid, Jordan. Dr. Nicolle has published over 100 books about warfare ranging from Roman times to the twentieth century, mostly as sole author. He also co-authored the Arab MiGs series of books, which covered the history of the Arab air forces at war with Israel from 1955 to 1973. Furthermore, he has appeared in several TV documentaries and has published numerous articles in the specialised press.

AIR VICE MARSHAL GABR ALI GABR

The late Air Vice Marshal Gabr Ali Gabr, PhD (EAF, ret.) served as pilot of a de Havilland Vampire fighter jet during the Suez War 1956. After concluding higher military education at the Air Warfare Institute in 1960, he served as Instructor in Air Tactics at the Air Warfare Institute in 1962–1964 and 1966–1967, as Staff Officer during the June 1967 War, Chief of Operational Training Branch in the period 1968–1973 and as Chief of Operations Group during the October 1973 War with Israel. After serving as Instructor in Art of Operations and as Chief of Air Force Chair at High War College from 1977 until 1982 and receiving his PhD at Nasser High Academy in 1989, he moved into writing and has published seven books and dozens of studies and articles on history of air warfare in Egypt and abroad.